Praise for *Generation Regeneration*

"Tony Cho is a placemaking visionary who combines spiritual enlightenment, a mission to preserve the environment, and decades of real estate development success. He knows how to dream and really manifest it. This book not only provides an excellent read but also offers insights and strategies we need to nurture both human happiness and planetary health. I'm a huge fan."

—**DAN BUETTNER,** #1 *New York Times* bestselling author of *The Blue Zones*, *National Geographic* fellow, and Emmy award–winning producer

"A complete and thorough introduction to the world of regenerative placemaking. Totally accessible to all, and a reminder that hope is far from lost for this planet as long as we have the right attitude, the right strategies, and the right people. Tony is one of them."

—**DONNA DEEGAN,** mayor of Jacksonville, Florida

"Cho offers practical solutions rooted in optimism and bold thinking, perfectly aligning with the challenges of our time. His work is essential reading for anyone who believes that our cities and communities can not only sustain us but regenerate us."

—**PETER H. DIAMANDIS,** *New York Times* bestselling author of *Abundance*

"Before picking up *Generation Regeneration*, I didn't realize how much our cities—and our lives—could be transformed by this revolutionary concept. Tony Cho breaks down complex ideas into something simple, accessible, and downright inspiring."

—**VISHEN LAKHIANI,** founder of Mindvalley and *New York Times* bestselling author of *The Code of the Extraordinary Mind*

GENERATION REGENERATION

Codesigning the Future of Cities Through Regenerative Placemaking

TONY CHO

Published by Greenleaf Book Group Press
Austin, Texas
www.gbgpress.com

Distributed by Greenleaf Book Group

For ordering information or special discounts for bulk purchases, please contact Greenleaf Book Group at PO Box 91869, Austin, TX 78709, 512.891.6100.

Design and composition by Greenleaf Book Group
Cover design by Greenleaf Book Group
Cover image © Adobe Stock / korkeng
Toolbox icon © Adobe Stock / Recoonde

Publisher's Cataloging-in-Publication data is available.

Print ISBN: 979-8-88645-508-3

eBook ISBN: 979-8-88645-509-0

To offset the number of trees consumed in the printing of our books, Greenleaf donates a portion of the proceeds from each printing to the Arbor Day Foundation. Greenleaf Book Group has replaced over 50,000 trees since 2007.

Printed in the United States of America on acid-free paper

26 27 28 29 30 31 32 33 10 9 8 7 6 5 4 3 2 1

First Edition

For Ximena.

CONTENTS

Foreword

As our world faces unprecedented environmental and social challenges, the way we engage with life has never been more critical. Regenerative placemaking is far more than traditional urban planning; it is a transformative practice that seeks to heal and restore a mutually beneficial relationship between people, nature, and the built environment. It also aims to go beyond healing to building an ongoing harmonizing capability between human beings and other life.

The ultimate goal of regeneration is a renewed understanding of how life uniquely thrives within each distinct socioecological system. By embracing this deeper understanding, we can guide design decisions while simultaneously empowering communities to care, connect, and contribute in meaningful ways. This creates a dynamic, coevolutionary relationship that sustains and evolves over time.

Through our work in the field as part of teams delivering regenerative projects, we've had the privilege of experiencing the transformative power of the unfolding process of regeneration, working with our places as acupuncture points to energize entire communities. This reinforced the connection among the people, Earth, the soil, the water, the social hubs, the animals, the food, the culture, the governance, and so much more.

We first connected with Tony over our shared passion for this work.

I (Hes) have long explored how regenerative placemaking can shape the future of our cities—through both scholarship and practice. I (Reed) have been in ongoing dialogue with Tony over the past five years about regenerative development and the centrality of "place" in guiding meaningful change. *Generation Regeneration* is a powerful and timely contribution that distills the potential of these ideas into something accessible to everyone. It shows that living regeneratively is not only the realm of practitioners and professionals—it is a meaningful, essential pathway for all of us who care about the future of our cities and communities

Generation Regeneration offers a blueprint for a future where cities and communities thrive in harmony with the natural world. The work detailed in this book is not merely theoretical; it draws from two decades of real-world experience in tackling the complex challenges of urban development. The strategies outlined here emphasize not just successes but also the difficult lessons learned along the way, highlighting the transparency and honesty required to navigate the complexities of regeneration.

While the practice of regenerating an ongoing healthy relationship with life—humans-to-humans and humans-to-nature—seems complex, once the door is opened to this way of being with life, we realize how simple and natural it is. In this book, regenerative placemaking isn't presented as a static theory but as a dynamic, living practice. The stories and experiences shared in these pages demonstrate the profound impact that regenerative strategies can have on the urban landscape and on the larger context of life our cities live within. Each example serves as a case study of how regenerative principles can reshape the fabric of our cities, making them more resilient, equitable, and attuned to the needs of both people and the planet.

As you read this book, you'll find a call to action that extends to policymakers, planners, architects, and everyday citizens. It challenges us all to think differently about how we build and engage with our surroundings. The future of urban development lies not in simply sustaining what we have but in regenerating and elevating it, creating communities that enrich both the environment and the human experience.

Humans have an essential role in nurturing the health of the ecological systems that support us. Just as ecosystems thrive through the balanced flow of resources, so too must our economic systems. This book is unapologetic about the vital role of money within a thriving whole. As a measure of exchange and value, it flows much like water, clean air, and ideas. Just as these resources can either enrich or deplete a place, money must also become part of the regenerative evolution of our cities and communities.

The most evolutionarily resilient ecosystems are those where humans, through thoughtful observation and active stewardship, contribute to increasing diversity. By embracing our role as coevolutionary participants, we can harmonize our needs with the life-sustaining processes, systems, and structures of the natural world.

As you journey through these pages, we hope you will find inspiration to become an active participant in this regenerative movement.

DR. DOMINIQUE HES, PhD
Adjunct research fellow, senior honorary fellow at the Melbourne School of Design, the University of Melbourne and Griffith University

BILL REED
Codeveloper of LEED, principal at Regenesis Group, managing director at the Place Fund, and leading practitioner in regenerative development

INTRODUCTION

Welcome to the Regenaissance

Inspire the mind; move the heart; change the world.
—Anonymous

Regenerative placemaking is for anyone who believes in creating thriving, resilient places where people and nature can flourish together. You don't need any particular qualification or title to make a difference. You just need the willingness to see, to care, and to take action. Consider this book your invitation.

In my experience, some of the most valued stakeholders in the world of regenerative placemaking are supportive community members and citizens who are willing to offer their time and energy to it. My aim with this book is not just to inspire hope for a better future but to equip you with the knowledge and tools needed to create it. This book is more than just a guide; it's an experiential toolkit and a blueprint, carefully crafted over two decades. I've codesigned and open-sourced this comprehensive framework with the team at Future of Cities® and global advisors, experts, and ordinary folks to ensure that it's accessible, understandable, and actionable for anyone who picks it up. My goal is to empower you with the practical wisdom and strategies necessary to make meaningful, lasting change in your community and beyond.

The book you're about to read draws on theory, but the content you'll discover is mainly built on experience. The language I'll use here is clear (I hope), and I've tried to humanize regenerative placemaking, bringing it to life. It's not my intention to put a dense textbook that feels laborious to read onto the dusty shelves of college libraries.

My vision for this movement is to bring everyone into the fold—no one left out, no one left behind. This isn't just a top-down approach; it's about humility, openness, and shared learning. I'm laying it all out here—my successes, yes, but also my failures, the missteps that taught me more than any victory ever could.

This book isn't just a collection of frameworks either, although you'll find plenty of those—from the groundbreaking ideas of Buckminster Fuller to the United Nations' (UN) Sustainable Development Goals and the pioneering work of the Regenesis Institute in regenerative development. It's a raw and real chronicle of two decades of hands-on experience. From helping to revitalize the Wynwood Arts District and cofounding the Magic City Innovation District in Little Haiti to reimagining Jacksonville's Phoenix Arts and Innovation District and nurturing the ChoZen Eco-Retreat and Center for Regenerative Living, every story in this book highlights both successes and lessons learned. My hope is that these experiences inspire you to forge your own regenerative path.

By the time you turn the last page, you won't just have a theoretical understanding of regenerative placemaking; you'll know how to take action, and, perhaps even more importantly, what pitfalls to avoid. You'll see the critical role you can play in cocreating a world that is not just sustainable but truly beautiful and regenerative, one worth passing on to future generations.

Why Caring Is Nonnegotiable

Once people know what regeneration means for them, the multiple issues it addresses, and what magic it creates, they realize that they can't afford *not* to care about it. We, as a species, cannot keep playing the role of Earth crisis managers indefinitely. Constantly focusing all our energy on putting out fires and expecting beauty and balance to take over as a result is futile. It's like trying to build a strong foundation by only patching cracks—an endless, exhausting cycle that never leads to lasting stability.

The real answer lies in preventing those human-induced fires from igniting in the first place and, even better, creating environments where life doesn't just survive but thrives. We need solutions that do more than just stop disasters; they should spark resilience and growth for everyone on this planet. It's time to look ahead, beyond sustainability, to a future where human evolution and global healing are the goals. That's where regeneration—or, more specifically, regenerative placemaking—comes in. Caring about regeneration isn't just important; it's absolutely essential because, if we don't start embracing regenerative principles in the way we build and live in our communities, we might not have much time left to care at all.

Unfortunately for me—and for any of my fellow environmental stewards, for that matter—sustainability as a subject isn't *sexy*. It's often framed in terms of sacrifice, like *reducing* our carbon footprint or *cutting back* on resource consumption. It's all about making *less* impact. In the team's early work at my company Metro 1, an impact-driven real estate brokerage, I observed the ineffectiveness of this approach firsthand. Back in 2005, as I and the Metro 1 team poured our hearts into building Miami's first green building resource center in Wynwood, we quickly realized that simply calling for people to do less harm wasn't enough to inspire real change. And that's putting it lightly. While the idea of building differently and integrating regenerative principles, such as innovative construction materials and renewable energy, into design was exciting to us, nobody else seemed to care.

It didn't catch on then, and it's an initiative that, in my eyes, is still failing to inspire action. We knew we had to go further and advocate for creating something better. We are now actively cocreating the Climate and Innovation HUB in Little Haiti—twenty years after our first attempt—this time with a much larger market of interest for regenerative thinking. It's a space designed to show, not just tell, what's possible when we integrate regeneration into real estate and urban development. This time, the market is more ready, the interest is greater, and the need is clearer than ever. My hope is that this hub serves as a living example of what's possible—demonstrating not only the environmental impact but also the long-term cost savings and community benefits that come with regenerative design and innovation.

Step by step, we continue to deepen our understanding of regeneration—exploring regenerative design, development, and placemaking—by actively cocreating and innovating these dynamic spaces, both online and in real life. Regenerative placemaking is not about minimizing our impact; it's about actively regenerating our communities and environments and, therefore, maximizing our impact. Because of this shift, our vision evolved from green buildings and sustainability to the bigger picture of cocreating places where people and the environment can thrive together.

Collaborating as a community to develop place-based solutions has been far more inspiring and impactful than working in silos. And more importantly, *ecosystems thinking* could save us all. Without it, we could face something more serious than a 2.5-degree rise in temperature.

Miami: Today's Atlantis?

The tale of Atlantis is unsettlingly relevant. In 9400 BCE, beyond the modern-day Strait of Gibraltar, there was a beautiful island called Atlantis. It was described by the Greek philosopher Plato as a utopia populated by a prosperous and technologically advanced society. The people of Atlantis achieved great feats of architecture, engineering, and governance. They had everything they could ever want or need. But, unfortunately, as time went on, the people of Atlantis became arrogant, greedy, and increasingly morally corrupt. They prioritized progress and growth over balance and sustainability, and, as a consequence, they took more and more from the nature surrounding them, unwilling to slow down or give back. Their resources were being exhausted. Their selfishness and individualism grew, and they lost their sense of community. The natural balance had been tipped, and the collaborative spirit that permeated their island—leading to their advances—was no longer present.

According to legend, one fateful day, the gods grew tired and angry with the Atlanteans' behavior and conspired to bring about their downfall. In a single day and night of misfortune, Atlantis was swallowed up whole by the sea, leaving behind nothing but a cautionary tale of pride and destruction.

While the fictional (or is it?) example of Atlantis may be grounded in Greek mythology, lessons can be learned from its fate, and precautions can be taken against our own collectively induced flooding. The sea level around Florida is rising steadily and substantially—as much as one inch every three years. And projections indicate that by 2070, the sea level in Florida will rise by just under a meter, sinking 1.7 million acres of land, forcing 906,000 people to relocate.[1] Many of those people will be forced to flee Miami for good. High tide flooding will be routine for everyone. This is a prophecy, a life-changing scientific prediction that seems to be coming true right now.

While Earth naturally goes through heating and cooling cycles, the last 100 years or so of industrialization have compressed what should have taken thousands of years into just a century, drastically heating the planet and accelerating global warming. According to NASA, this warming rate has not been witnessed on Earth for the past 10,000 years.[2] We're no longer just witnessing these changes; we're living through them, and they're hitting us faster and harder as time progresses.

This is my home, but just because you might not live in Florida doesn't mean you're removed from the effects of climate change. If you're reading these words from planet Earth, we're in this together, my friends. According to projections from Cornell University, by

2100, two billion people will be displaced from their homes due to shrinking land masses, creating a massive wave of climate refugees.[3]

That's by flooding alone. It doesn't include the numbers for hurricanes, heatwaves, wildfires, droughts, famine, and disease. Resilience in the face of this incoming climate change is essential for our survival. This is why I've committed my life's work to regenerative placemaking—not just as a concept or theory but as a practical framework that I actively apply across all my projects. My goal is to demonstrate what's possible, to show that we can go beyond minimizing harm and instead create places that enhance everyone's quality of life.

Atlantis doesn't have to be our future. We have the power to change course. As a species, we must evolve, to prototype and implement the many solutions that already exist.

Regenerative Placemaking: A Roadmap to a Thriving Future

I, as well as my amazing team at Future of Cities® and most other committed regenerative placemakers, do believe there is a solution to avoid further catastrophe. *Our mission is to protect and transform the lives of one billion people by pioneering innovations in the built environment through the practice of regenerative placemaking.*

We haven't come this far to see our homes, some of which we built with our bare hands, sink in front of us. The possible solutions to climate change involve cross-sector collaboration at an unprecedented scale from diversified perspectives within the world of regenerative placemaking. And what may come as a surprise to some is that the results of this work offer vast economic benefits.

Regenerative placemaking is so much more than making what was once ugly beautiful, what was once dangerous safe. It is the key to community and climate resilience, as well as the new life breathed into decaying and abandoned areas. We have witnessed communities shift from bulk-buying food from distant lands to growing their own produce locally, from burning fossil fuels to embracing renewable energy sources like solar panels on their roofs, and from relying on cars to using amazingly clean, punctual, and comfortable public transportation. We've seen entire neighborhoods and natural environments go from merely surviving to thriving. And communities are cocreating pockets of *protopia* all around the world.

Change starts with the individual. We must do the inner work needed to *feel* regenerative from within before extending to the collective—a key concept in regenerative development theory. However, luckily, positive transformation can begin at any entry point within a system. While some may start by focusing on improving urban life quality or reducing environmental impacts, a deeper understanding of interconnected natural systems often follows, highlighting that isolated actions (like green buildings) are only a part of the solution. The broader whole requires a deep healing process.

Even though change starts with the individual, nature didn't design us to be individualists; we do best when we work together, support each other, and stay united. Blaming, dividing, and operating without a sense of unity only weaken our potential for meaningful impact. Regeneration is only possible when we work together. It doesn't mean we have to agree on everything, but it does require a shared commitment to core values—like clean air and water and healthy soil—that benefits us all.

Kick off of Regenaissance with Buckminster Fuller Institute and Design Science Studio in collaboration with Future of Cities® during Miami Art Basel 2021 in Magic City Innovation District; Photo source: Zahra Shohadaee

Regenerative placemaking holds the power to address some of the most pressing challenges facing our world. If left unchecked, these challenges could lead to our collective downfall. It's a reminder that true progress and thriving are only possible when everyone and everything around us also has the chance to rise. We must recognize that our success is intertwined with the well-being of our communities and the environment.

Beyond climate change, regenerative placemaking supports communities to understand and develop solutions for problems they may be facing. These problems include the following:

- Social inequality: Regenerative processes can support marginalized groups to identify their potential and gain access to resources, spaces, and systems that facilitate participation in social change and the general decision-making processes of their city and community.

- Poverty: Regenerative placemaking develops the capacity to create jobs for local people and supports small, sustainable businesses, boosting local economies, creating circular economies, reducing economic inequalities, and encouraging community resilience and self-sufficiency.
- Urban and suburban sprawl: Regenerative placemaking by design focuses on accessible public transportation, smart density, land conservation, less reliance on cars, and lower carbon footprints and pollution.
- Cultural misappropriation and loss: Regenerative placemaking can address the preservation of local cultures and traditions by integrating them into the development process and maintaining cultural identity and heritage through celebrating and protecting the story of place.
- Mental and physical illness: Green spaces and regenerative urban design play a vital role in nurturing healthy ecosystems—both human and natural. They enhance air and water quality, encourage physical activity, and provide mental health benefits, such as reducing stress and building social connections. Together, these elements contribute significantly to overall public health.
- Famine: Regenerative placemaking incorporates regenerative agriculture and sustainable farming practices, ensuring local

food production, reducing reliance on long supply chains, and increasing food security.

- Water scarcity and surplus: Implementing sustainable water practices like harvesting rainwater, recycling greywater, and creating natural water filtration systems reduces the risk of water scarcity, droughts, and floods.

The way to safeguard our future—and that of our children and grandchildren—lies in embracing regenerative placemaking as a process of appropriate and evolving solutions. We've taken more from the planet and our communities than we've given back, and we've made significant mistakes along the way. But it's not too late.

As a relatively young species, we're just beginning to realize the consequences of our actions. Now that we're aware of the ecological and social damage we've caused, it's time to reverse the harm. This means fundamentally changing how we live and rekindling a true sense of community. By doing so, we can begin to heal the planet and ourselves.

My team and I are doing this in Florida right now. Yes, it's a vulnerable area of the United States, and many people ask, "Why not just sell everything and leave?" All my roots are here in Florida, and my answer is simple: If not me, then who? This is my native land, and it demands that we develop steadfast dedication to being part of its healing. Through some of our Future of Cities® demonstration projects across Florida—and now extending across the Atlantic to Portugal with the FOC Golden Visa Fund—we're taking bold steps to build resilient communities and honor our responsibility to protect and regenerate the land.

We're leading a revolution in our Climate and Innovation HUB in Little Haiti, Miami. The Climate and Innovation HUB is now an open house that's all about tackling climate change and its effects in the city and in our state. We're laser-focused on rapidly discovering and prototyping new solutions to help Florida survive, stay sustainable, and thrive as it regenerates. Collaboration is a key part of what we do, bringing together everyone—government officials, business people, investors, brokers, researchers, and nonprofits—to work on solutions and share ideas. Plus, and perhaps even more importantly, we get local communities involved—raising awareness and encouraging people to take part in climate action and regenerative placemaking.

Protecting life on Earth doesn't have to be all work and no play either. People have an amazing time making an impact through regenerative placemaking. A brilliant local group of regenerative placemakers recently cocreated a regenerative project for the well-being of sea life and humans alike.

Miami Beach, with its beautiful but endangered coral reefs, has faced significant damage over the years due to a combination of environmental stresses and human activities. About 70% of the life of the reef has died off through overfishing, pollution, and climate change.[4] That's scary for fish. But what's scarier for us is that we are supposed to eat the fish that are quickly vanishing, and healthy coastlines can absorb 97% of wave energy from storms and hurricanes.[5] Do the math; it equals food scarcity and flooding.

A small yet passionate group of regenerative placemakers asked themselves what they could do about it, and the local artists, creatives, architects, and scientists had an idea. Leading this group of underwater explorers is Ximena Caminos, the founder of the REEFLINE project. This ambitious project will create an underwater art gallery that doubles as a vital habitat for sea creatures and a protective barrier for the land, promoting coastal resilience. After years of ideation and pitching, the concept recently received funding and is now becoming a reality, with the first piece of the reef slated for installation in 2025. The team includes an all-star lineup of architects from the international Office for Metropolitan Architecture, marine biologists, and other experts who continue to do incredible work. Led by Caminos and her design team, they've shown remarkable creativity, innovation, and tenacity—navigating approvals from the Army Corps of Engineers, the Miami Beach government, and various grant providers to bring this vision to life.

Concrete Coral—Leandro Erlich, 2025; This submerged reincarnation of Leandro Erlich's celebrated Order of Importance (2019) will install a life-size traffic jam of twenty-two concrete car modules for sealife to colonize. Symbolically reversing the effect our actual cars have on the environment, these sculptures will contribute to restoring damaged natural habitats.

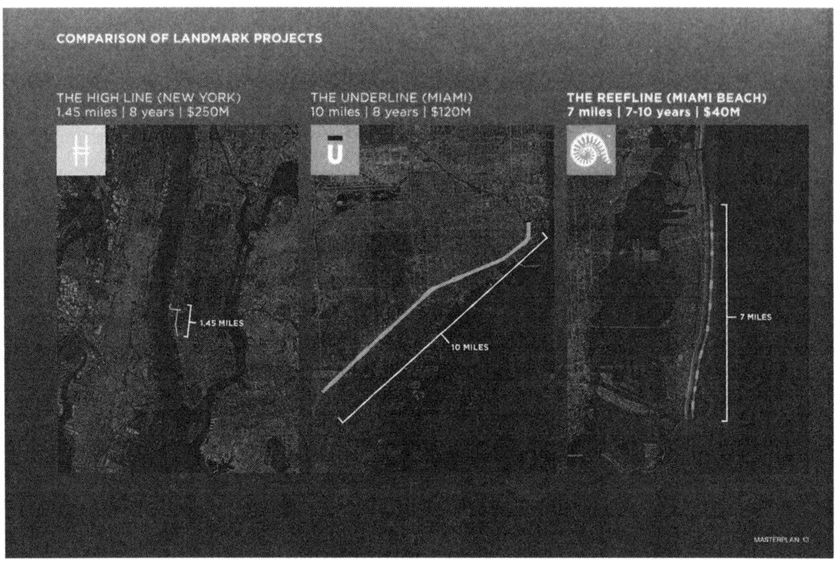

Masterplan for The REEFLINE & OMA in Miami Beach 2025; "NY Has the High Line, Miami Has The Underline, and Soon, Miami Beach will Have the REEFLINE" — *The New York Times*

Our team at Future of Cities® has been learning from the REEFLINE as they've opened up their process through local and global salons, workshops, and roundtable dialogues—offering a rare behind-the-scenes view into the development of their underwater sculpture park and hybrid reef. We're honored to feature the 3D-printed Miami Reef Star prototypes at the Climate and Innovation HUB and open our doors to all who are curious to dive deeper into the project and explore further their innovations in material science, ocean literacy, and the future of regenerative design.

Our doors will always be open to everyone involved. The REEFLINE is profoundly inspiring, and the dedication of those working on it is truly remarkable. Through extensive testing, research, and cross-sector collaboration, the waters just off Miami

Beach have now become a destination for eco-tourism, cultural innovation, and marine education—made possible by the love, perseverance, and unwavering dedication of the REEFLINE team.

Opportunity Zones and the Financial Benefits of Placemaking

When done right, regenerative placemaking doesn't just transform neighborhoods; it creates real, lasting value for communities and investors alike. Financial return is often a taboo subject in social impact spaces, but for us, it's a sign that doing the right thing can also be sustainable. My goal has always been to prove that creativity, collaboration, and care can generate outsized returns—socially, environmentally, and financially. The Opportunity Zone (OZ) program, introduced in 2017, opened up a powerful pathway for this. With major tax incentives for investments in underfunded communities, OZs were designed to stimulate economic revitalization, although the results have varied depending on the intent behind the projects.

At Future of Cities®, we've been using OZs not to extract but to regenerate. Our project in Jacksonville, Florida, is a living lab for inclusive growth, where we are restoring historic buildings through adaptive reuse, nurturing entrepreneurial talent, and codesigning the district with the local community. We seek to prototype and introduce models and concepts like profit participation and fractional ownership, allowing community members, artists, and small business owners to share in the financial upside. Technologies like blockchain and DAOs (decentralized autonomous organizations) offer even more promise: enabling transparency, distributed decision-making, and a broader sense of ownership over how funds flow and who benefits. It's not just about building new spaces; it's about building with people, not over them.

To make regenerative placemaking viable, we combine OZs with tools like green bonds, new market tax credits, and PACE financing, creating a stable foundation that supports affordability and inclusion. Our experience shows that when incentives are layered strategically and aligned with community values, money becomes a tool for healing rather than for displacement, one that elevates people and protects culture. Regeneration isn't just an ideal. It's an investment—one that pays off when we lead with purpose.

In July 2025, the nonprofit arm of PHX-JAX, Friends of PHX-JAX, received an appropriation from the state of Florida for US$350,000 to launch Bloom Labs, an after-school program for students focused on urban agriculture, entrepreneurship, and innovation. This was a bipartisan effort from Senator Tracie Davis and her Republican counterpart Congressman Wyman Duggan, supported by Governor Ron DeSantis, demonstrating that when doing regenerative placemaking, all parties can and do come together for a shared purpose.

Rewind to 2005: When we started codesigning Wynwood, there were no tax benefits, no OZs, and no structured frameworks for regeneration. What we were doing at that time was "creative placemaking"—bringing artists, galleries, and cultural energy into a forgotten district and transforming it into something magnetic. And while the impact was undeniably powerful, the financial rewards didn't always reach the people who helped make it happen—the small business owners, local artists, and early collaborators who poured their hearts into the neighborhood.

Wynwood 2018 vs. Wynwood 2025; Photo Source: Metro 1 Commercial by Bruno Vitale

In hindsight, Wynwood wasn't regenerative. Not yet. It lacked the long-term equity structures, affordability mechanisms, and financial tools we now understand are essential for true regeneration.

Wynwood taught me a lot—and that's exactly why, when OZs were introduced years later, I saw them not just as tax incentives, but as a second chance—a chance to do it better, to design projects that don't just revitalize a place but honor and sustain the people within it.

The Story of the Wynwood Arts District

You wouldn't know it now, but Wynwood Arts District was, in the early 2000s, a collection of mostly abandoned shoe and garment warehouses dotted along old train tracks. It was also predominantly home to very underrepresented Puerto Rican residents. Known as

"El Barrio," the neighborhood received little attention from visitors and was unsafe and underinvested in by businesses and from the city. It was a largely forgotten landscape that underwent some heavy problems with crime.

The area, once known as the epicenter of 1980s unrest and home to the Miami Riots, seemed to be on a trajectory toward rapid decline. But where others saw grit and danger, I saw untapped potential. With a belief in the transformative power of the potential of place, I took a chance. I recognized that, with the right vision and investment, this area could be revitalized and reimagined into something truly vibrant and thriving.

Wynwood 2013; Photo Source: Ben Coppelman

In real estate, it's often said: *Location, location, location.* Wynwood, nestled between the airport and Miami Beach, just north of downtown and south of the Design District, had that prime location. *What an opportunity.* But almost no one saw it—except a few optimists like myself, Tony Goldman, David Lombardi, the Rubell and Margulies families, and a whole host of artists, creatives, and gallerists. We read the tea leaves and believed that a magical transformation could be on the horizon if we could bring together the right stakeholders. We dove in and never looked back.

The RC Cola Plant circa 2011; Photo Source: Metro 1 Commercial by Bruno Vitale

In 2005, I leased my first office on the railroad tracks in Wynwood and founded my first company, which would later become Metro 1 Commercial. The amazingly talented musician Lenny Kravitz had a studio right next door, and there were a lot of underground artists starting to paint the neighborhood. From the start, Metro 1 wasn't the biggest player on the scene, but it was the first to do many things—focusing on the urban core and shaping place through thoughtful tenant curation. Over time, it became an award-winning brokerage and a driving force behind Wynwood's transformation, eventually growing into one of the most influential real estate brokerage and investment firms in Miami. In 2011, I was honored as the Young Leader of the Year by the Urban Land Institute at their Vision Awards, a recognition that underscored Metro 1's role in leading the way.

Wynwood 2013; Photo Source: Ben Coppelman

During that time, we cross-collaborated extensively with street artists, landlords, locals, creatives, and designers, and as the firm grew, the Wynwood Arts Association emerged with a shared commitment to transform and revitalize the area. One of our most brilliant agents, Nina Arias, an artist herself, cofounded it, and Metro 1 Commercial actively participated in and supported its efforts.

In Nina's words, "In the early 2000s, Wynwood was a raw, untapped canvas—a gritty industrial neighborhood on the brink of transformation. As a young independent curator in my twenties, I recognized its potential and opened my first contemporary gallery, Rocket Projects, joining a wave of visionary artists, gallerists, and collectors with a shared passion of creating a new art community. The energy was electric and contagious and a burgeoning art scene was being born. Together with Brook Dorsch, Locust Projects, Don and Mera Rubell, and Marty Margulies, we helped lay the early foundation for what would become the Wynwood Art District. I played a key role in designing the Wynwood Art District Association, helping to shape the vision, infrastructure, and cultural identity that included an exciting monthly gallery walk, a printed map, and neighborhood street banners that would define the neighborhood and its community.

"In 2005, I joined forces with Tony in real estate at Metro 1 Properties and started cocreating dynamic art events that brought people together, igniting Wynwood's emerging cultural scene. Those early days were a thrilling experiment in creativity and urban reinvention—a testament to the power of art to transform not just a place, but the people who experience it. Tony was already on the right track with his visionary approach to blending real estate with creative placemaking, a philosophy that would continue to shape communities far beyond Wynwood."

The Wynwood Art District Association evolved rapidly, and I ended up serving as an original board member for the Wynwood Business Improvement District. Together, we worked to clean up the area and establish a unique identity. Artists painted extraordinary murals on warehouse walls, transforming them into captivating canvases that soon attracted even more artists, investors, developers, brokers, and landlords from across the country—and eventually the world.

The Wynwood Building by Goldman Properties avant-garde business park with an art-filled exterior housing retail & office space circa 2018; Photo Source: Metro 1 Commercial

The once-empty streets are now vibrant and alive, lined with bustling cafés and restaurants. Regularly organized art walks invite the public to explore the stunning murals and galleries, fully immersing themselves in what has become the epicenter of street art and creative culture in Miami. In regenerative thinking, we talk about vibrancy and aliveness—and Wynwood embodies that spirit. It's definitely alive.

As you can imagine, it's become a place where everyone wants to live. As Wynwood came about before the rise of regenerative principles, it was developed so well—too well. It became a victim of its own success. Many of the original artists and creatives that brought it out of the shadows can no longer afford to live and work there. But it does, however, stand as a powerful example of what's possible with a vision, a strong sense of community, and collaboration.

So, the financial returns for most of the top investors were very high, while the results for locals were . . . mixed. Many small businesses have now been priced out as the area has transformed, a reality I witnessed firsthand and one that motivated us to later launch Future of Cities®.

It became clear that there was more to be learned and more to be done beyond revitalizing neglected areas for the sole profit of the biggest investors. There was a deeper impact to be made—an impact that went beyond financial gain to truly support those with limited resources. At just 26, I had the incredible opportunity to build with a vibrant community and witness Wynwood's evolution fueled by its homegrown creativity. The experience was unforgettable and profoundly shaped by an understanding of the power we have to transform our neighborhoods through art and innovation.

Asking the Right Questions

My team and I at Future of Cities® went on to question how we could bring financial freedom not only to those involved in the development of underrepresented urban areas but to the communities that lived there too.

Was there a way to develop that didn't result in gentrification and unaffordable housing? Can we not only revitalize places but also create lasting opportunities for future generations? How could we support local ecosystems both in the built environment and the natural environment? Ultimately, how can we protect our neighborhoods and our planet simultaneously?

The answers to these questions became the foundation for our team at Future of Cities®, inspiring us to embed holistic success through regenerative placemaking grounded in three core pillars: community, nature, and culture. That was the moment when we went from bringing places back to life to cocreating regenerative places that would give life to people and the planet. We went from creative placemaking pioneers to regenerative placemaking innovators. In other words, we became part of the regenaissance, and so can you!

Regenerative placemaking is rewarding on every level, and there is a lot to be gained from real estate development in this arena. But it's more than a career for me; it's a calling that aligns with both my personal and collective purpose.

According to the Regenesis Group, a leading authority in regenerative development and the institution that led my regenerative practitioners studies, vocation transcends the idea of a job; it's a deep, guiding purpose that connects individuals with the needs of their community and the environment.[6]

In regenerative development, responding to this vocation isn't about following orders but about answering a calling that speaks to the essence of who we are and what we can contribute. It's about aligning our actions with the broader potential of the places we engage with, creating a meaningful impact that goes beyond mere sustainability. When a community embraces this potential of place, it doesn't just rejuvenate the built environment; it revitalizes its collective identity and purpose. This mutual regeneration is at the heart of truly transformative regenerative placemaking projects, where both the place and its people evolve together, creating a brighter future.

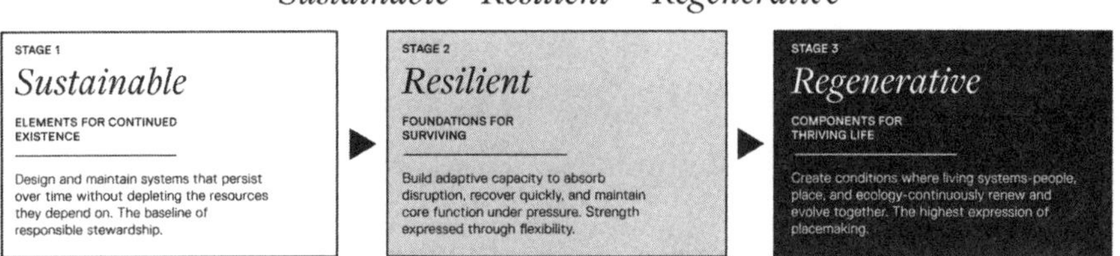

Sustainability vs Resilience vs Regeneration — as defined by the Regenesis Institute

As you continue to explore the concept of regenerative placemaking in this book, remember that it's not just about creating beautiful spaces—it's about accepting a new, meaningful potential of place, a new story, one that will be told by both current and future generations. For me, for you, and for all of us, regenerative placemaking is a way of living, one that shines a light on the interconnectedness of everything. It highlights the important role all of us have as caretakers and stewards of the land, and it urges us to think more broadly and holistically about our impact.

CHAPTER 1

Regenerative Placemaking—What on Earth Is It?

Healthy cities are those that are continually being made and remade by their inhabitants.

—JANE JACOBS

Anyone on my team will tell you: When they're asked what they do for a living, and they say that they're *regenerative placemakers*, most people respond with a brief, curious silence. Regenerative placemakers come to appreciate this pause; it means that our new friend or acquaintance is probably unfamiliar with the role, and we have the chance to introduce them to something truly transformative. We're well aware it isn't fully mainstream just yet, but we have unshakable faith that it will be.

The challenge is to describe what regenerative placemaking is in a simple yet exciting way (well, I think it's very exciting). Explaining regenerative placemaking allows us to go beyond dictionary definitions, sharing stories of places and communities that have risen from decline to embrace regeneration, evolution, artistry, connection, and resilience. We can talk about how marginalized cultures have been uplifted, how pollution has given way to green systems, and how mindful investments have yielded returns.

I personally have only been actively identifying with the term *regenerative placemaker* since 2020, but it's an identity that my team and I at Future of Cities® have all been journeying toward for decades. In reality, regenerative placemaking is a natural evolution, merging our experiences as creative placemakers, real estate investors, brokers, anthropologists, environmentalists, and developers.

My journey as a regenerative placemaker hasn't been without its lessons. I saw the meteoric rise of places like Wynwood and Little Haiti, with skyrocketing monetary and cultural value. But with this success came the unfortunate displacement of lower-income creatives and small businesses who could no longer afford to live in the area they had helped build. This experience has underscored the importance of a regenerative approach—one that balances growth with inclusivity, cultural preservation, and environmental respect.

Regenerative placemaking isn't just a lofty ideal; it's a catalyst for deep, meaningful change. It's an ongoing process. It drives transformation that reaches beyond visible results, creating social cohesion, economic opportunity, and environmental stewardship. I've been fortunate to witness neglected urban areas becoming vibrant cultural hubs and to see grassroots projects offering renewed hope and identity to underrepresented communities. All this progress is underpinned by a commitment to improve life for everyone.

Fundamentally, regenerative placemaking is all about creating places that positively serve people and protect the natural world they're part of. But that's just scratching the surface. The term encompasses so much more than what is commonly understood, yet it has long been misinterpreted in ways that totally miss the mark and therefore potentially put people off getting involved with it.

Before I introduce the definition of *regenerative placemaking* written by our team at Future of Cities®, I want to dispel some common misconceptions. By understanding what regenerative placemaking is *not*, we can gain a clearer and deeper understanding of what it truly is. Let's debunk these myths and chart the way for a comprehensive understanding of this revolutionary concept.

What It Isn't

The first misconception is that people think regenerative placemaking is just about green buildings—ecofriendly structures with a living wall, moss chandeliers, and a dubious scattering of vegetable and flower patches. However, in reality, although green building practices are definitely part of it, regenerative placemaking is much broader. Its environmental impacts go beyond planting more greenery for restoration and into exploring revolutionary ecofriendly power sources, water filtration systems, and rewilding initiatives that reverse damage caused by climate change. Moreover, as mentioned earlier, it includes social, cultural, and economic dimensions, focusing on creating holistic environments that enhance the coevolution of both human consciousness and the planet.

Some think regenerative placemaking is all about rich, middle-aged white businessmen taking over poor neighborhoods and colonizing them for profit. However, when done correctly, it is the opposite of top-down urban development. It is a community-led and inclusive process that prioritizes the needs and voices of local residents, especially those who are often marginalized. I see it like this: These local people may not have the financial means to influence projects, but they have a wealth of knowledge when it comes to understanding the place. The goal is to empower the community to actively participate in shaping their environment, ensuring that improvements benefit everyone, particularly the existing residents. Regenerative placemaking fights hard to avoid gentrification, displacement, culture extinction, and placetaking and prioritizes win–win results.

Many equate regenerative placemaking directly with sustainability and sustainable development. But it goes way beyond sustainability. While sustainable development aims to minimize harm, regenerative placemaking seeks to rejuvenate ecosystems and communities and set them up for an ongoing process of evolution. Sustainability is necessary but not enough. Sustainability is the bare minimum we need to get by and not exterminate ourselves. Think about sustainability as a life support machine, built to fight off death. It works for a while but doesn't make life worth living. Regenerative placemaking honors being alive and making that life worth living by actively enhancing the vitality of natural and human systems. It involves creating spaces that thrive, where the environment, culture, and economy flourish in harmony. This approach creates resilience, innovation, and holistic well-being, ensuring that future generations inherit a world that is not just preserved but improved.

Some people believe regenerative placemaking is a total city center concept that ignores the needs of those who live outside the heart of a city. In fact, it is applicable in rural, suburban, and urban contexts. It involves understanding and enhancing the unique characteristics and needs of any place, whether it's a city, neighborhood, small town, or rural landscape.

Regenerative placemaking is often thought to be designed and implemented solely by experts and millionaire investors, but everybody has a role. Effective regenerative placemaking is highly participatory, involving local communities in any and all big decisions and drawing on the wisdom and needs of the people living in those places. Yes, urban planners, landscape architects, developers, investors, and policymakers are 100% needed. But so are community leaders, students, activists, and the people who bring culture to a place to make that place home. As Ethan Kent of PlacemakingX described, "Public spaces and the people that protect and improve them transcend borders and reflect all stripes of humanity."[1] In regenerative placemaking, everyone is celebrated for their contributions, and we're in it together.

People often assume regenerative placemaking is prohibitively costly and not financially feasible. While the initial investments may seem higher risk due to the newness of the concept, the potential benefits are significant for all involved. Regenerative placemaking focuses on the coevolution of whole systems, considering impacts far beyond just financial gain. When we talk about capital here, it's essential to include various forms of capital—social, human, cultural, and natural. This approach can indeed be highly profitable for all stakeholders, provided everyone works together toward a common vision and higher goals. Alongside offering an economic return on investment, regenerative placemaking can also reduce operational costs, enhance property values, and promote circular economies. There's huge potential for blockchain, DAOs, and tokenized real estate integration within this framework, as well as increased financial resilience for those living in a regenerated community.

Some people assume that regenerative placemaking is mainly concerned with beautifying spaces. Although aesthetics are important, and everyone would prefer to live in beautiful environments, the core of regenerative placemaking is about functionality as well. It improves places based on the needs and desires of the people who live there, as well as supporting the natural environment surrounding them. This is a surefire way to start enhancing people's quality of life. It prioritizes ecological and human health alongside visual appeal.

What It Is

Historically, defining *regenerative placemaking* has been a feat that is more complex than you'd think.

When I first began to explore regenerative placemaking, I quickly realized that nobody could explain it without using an overly complex sentence (and usually an Oxford dictionary). So, when I launched Future of Cities®—a regenerative placemaking, development, and thought leadership platform—I collaborated with brilliant colleagues like Dr. Dominique Hes, Bill Reed, Ethan Kent, and others in the field of regeneration to codesign and open-source the concept. Together at Future of Cities®, we came up with this definition:

> Regenerative placemaking is about bringing people together to cocreate spaces that unlock the potential of people, place, and nature, helping communities and the environment thrive.

Ta-dah! We tried and tested that definition with everyone—from billionaire investors to humble community members—and it seems to resonate universally. That's the essence of regenerative placemaking: creating a level playing field where everyone understands what's happening and why.

Regenerative placemaking goes beyond understanding ideas; it invites us to feel and explore them personally, asking ourselves, *Does this resonate with me?* This reflection helps each person connect with the ideas in their own way, making it easier to adopt, question, or redefine them according to their unique perspective.

To make regenerative placemaking truly effective, we need to embody these ideas, bringing them to life within us. It's about uniting people around a clear, shared vision and working collaboratively to achieve it. Of course, this is a simplistic definition of a much wider concept that goes very deep down environmental and anthropological rabbit holes. But for the sake of clarity, I've found shortening the definition to be quite useful in the work we do.

In the following Venn diagram, you'll notice that regenerative placemaking consistently revolves around three core concepts: community, nature, and culture. These elements are central to our understanding and practice of regenerative placemaking at Future of Cities® and make up our three-pillar foundation centering on all life forms.

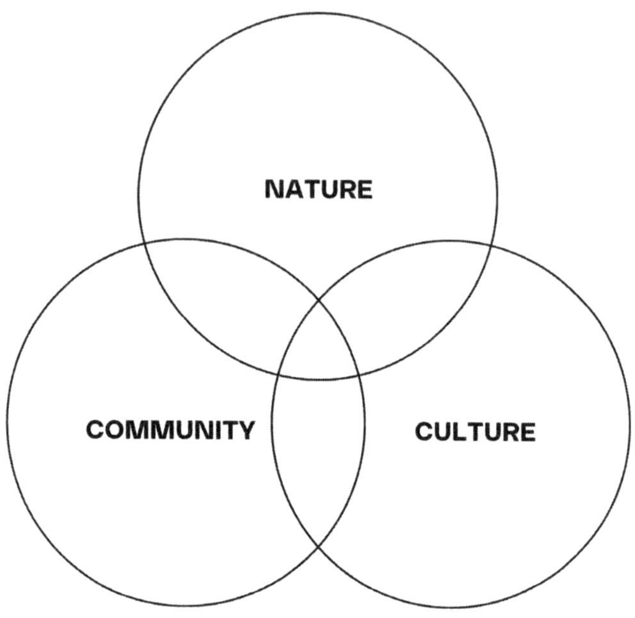

The three pillars of Regenerative Placemaking with Future of Cities® 2023

Yet regenerative placemaking isn't just about ecology, community, or culture in isolation; rather, it's about integrating all these elements into a cohesive whole. It's about seeing everything and everyone as completely interconnected.

Think of it as being like human anatomy: To achieve true health, we can't just treat one part of the body—a bone, a limb, or an organ—without considering the entire system holistically. Similarly, in our communities, we must consider all aspects together to achieve overall success, health, happiness, and economic abundance. Focusing on just one element in isolation risks causing breakdowns in other areas.

Regenerative placemaking includes considering the land, food, farming, community, businesses, governance, cultural identities, and community well-being as one. By recognizing how these elements are deeply linked, we can better appreciate the mutually beneficial value regenerative placemaking projects can provide within these systems. The ripple effects are endless, as are the possibilities.

The Potential of Place

To truly put regenerative placemaking into action, we must understand and harness the potential of each place. When I say *place* in regenerative placemaking, I don't just mean the built land. *Place* encompasses every aspect of life that makes it what it is—nature, culture, people, the invisible energies that connect relationships, and realms we may only barely understand. Partnering with a place as a regenerative placemaker means seeing it as a living entity, not just as land to be managed. It's about nurturing a relationship between the land and the community.

Indigenous perspectives, such as those of California's Miwok people, remind us that land loses its vitality when humans stop interacting with it meaningfully. As Kat Anderson explores in *Tending the Wild*, Indigenous communities cultivated this connection by tending and caring for the land, maintaining its health and richness.[2] Regenerative placemaking restores this kind of connection, bringing renewed meaning and purpose to entire living systems.

Regenerative design begins with potential—the hidden potential that can lead to future success—rather than focusing solely on problems. This approach involves evolving the system's capacity to make unique contributions to larger systems. For example, a neighborhood park has the potential to contribute to the community's health and well-being. This begins from the potential of shifting the focus from what's broken to what is possible, opening up new creative energies.

For instance, stormwater runoff is typically seen as a problem, prompting cities to invest heavily in infrastructure to get rid of it. However, many communities worldwide are beginning to recognize stormwater as a valuable resource, transforming it through water collection systems that enrich the environment. A really good example of this is Singapore's Active, Beautiful, Clean Waters Programme, which has successfully turned urban waterways into naturalized rivers and vibrant public spaces.[3]

This potential of place springs from the integration of human and natural systems in specific locations, advocating for *tailored* innovations and solutions rather than one-size-fits-all approaches. Different places require different regenerative placemaking strategies. The needs of a North African community differ vastly from those of a city like Pittsburgh, not just because of their environmental conditions but also because of their distinct cultural and economic contexts.

René Dubos, the environmental scientist who coined the phrase "think globally; act locally," argued that environmental issues—and any issue at all, for that matter—must be addressed within their specific physical, climatic, and cultural contexts. He saw each place as "the continuously evolving expression of a highly complex set of forces."[4] I agree. In other words, places are dynamic, and understanding them is not simple.

To grasp a place's uniqueness, we must recognize how it is interconnected, how it functions, and what gives it its special character. The framework I work with most advocates for the community, nature, and culture approach. To design with the potential of place in mind, we must consider three key questions[5] that help us understand a place as a living system:

How big is "here"?

This question,[6] coined by ecological artists Helen and Newton Harrison, allows you to identify the broader systems that your project will impact. Recognize that every project is part of a larger context, whether it's a building within a neighborhood or a neighborhood within a city.

How does "here" work?

Look at how natural features, local plant and animal life, and human activities interact in the area. This understanding allows for designs that fit seamlessly into the environment. Go back in time and understand how this place was formed; geological forces that molded the land continue to influence us today, shaping both our environment and our cultures in ways that are still unfolding.

What kind of "here" is this?

Engage with local people to learn what they love about their place. Their insights help ensure that the design respects and preserves the area's unique character.

These questions are broad but essential for understanding the characteristics and dynamics of a place. By exploring them, we can design projects that not only respect but also enhance local identity and special qualities, encouraging deeper connections between people and their environment.

As you can see, regenerative placemaking goes far beyond sustainability; it isn't about maintaining the status quo but about building upon what already exists, allowing a place to truly shine and realize its full potential. It's time the human race moved beyond the limited goal of sustainability. We must create environments and communities that inspire healing and revival for both people and the planet while building resilience against degeneration and challenges. This requires focusing on potential, not just problems. Our cities should be places of livability and joy, not just survival. By learning from our past mistakes, we can chart the way toward a brighter future.

A Shared Protopia

Have you ever heard about the concept of protopia? And no, this isn't a typo. It's different from utopia. Protopia and utopia offer two contrasting visions for a better, happier, more peaceful, regenerative society. Utopia is a perfect, heaven-like state where nothing ever goes wrong or challenges us. In a nutshell, it's a delusional goal, one that doesn't acknowledge the natural balance and constant flux of life. Protopia, a term coined by futurist Kevin Kelly, is about incremental improvement. It aligns with the Japanese concept of *kaizen*, "incremental improvement," representing a culture that is constantly evolving and getting closer to what is best for all life on Earth—even if it never reaches perfection.[7] Regenerative placemaking seeks to make this more achievable goal a reality. After all, perfection and utopia are, just like beauty, in the eye of the beholder.

If I were to ask you to define your idea of protopia, what would that be? Think about a place that serves not just you but everyone around you, a place where everyone wins and feels at home. It doesn't have to be perfect, and you can rest in the knowledge that it will evolve over time to get better and better.

What would your ideal home, city, or neighborhood look and feel like? What kind of life would you live? Now I don't know exactly what you thought of, but I'd bet you didn't say you wanted to live in a place with lots of concrete and pollution with no green space. I'd also guess that your protopia didn't consist of you being completely isolated from people, nor was your vision devoid of any and all culture or customs that meant something to you and your ancestors.

Although all of our individual versions of protopia are slightly different at first glance, they do share some things in common. What they share in common, and the patterns between each individual protopia, go on to create a *collective* protopia. Protopia, unlike utopia, can be shared among us all. Utopias are made in the eyes of the individual, and protopia in the eyes of the collective.

There are three, ever-present pillars of protopia that just so happen to be the same pillars I use as a framework at Future of Cities®: community, nature, and culture.

Community has been at the core of my life since birth. I was raised in an interfaith intentional ashram founded by my grandmother, Ma Jaya Sati Bhagavati, who adopted me as her own. She instilled some very important values in me like the importance of service and spiritual practice.

Growing up there felt like living in a bubble—protected from the outside world in many ways during my formative years. The ashram provided a humble yet deeply enriching childhood: I slept on the temple floor alongside my grandmother's disciples, played by the water with friends, and took on my share of responsibilities—washing hundreds of dishes and chopping garden vegetables for communal meals. Living in this close-knit environment instilled in me the values of kindness, cooperation, and service.

When I went to Chicago for college, I felt an overwhelming sense of isolation. For the first time, I truly understood how deeply community was woven into my DNA. While some aspects of ashram life didn't resonate with me, the sense of belonging and shared purpose did. My grandmother built her own version of utopia from the ground up, and it continues to evolve today. In fact, I purchased some of the ashram's land to protect it from reckless development and reimagine it as a nature-based ecoretreat and bioregional hub—ChoZen Eco-Retreat and Center for Regenerative Living.

ChoZen: A Bioregional Hub for Regeneration

While I believe that utopias, in the idealistic sense, are unattainable, ChoZen represents a modern evolution of community—one that exists in harmony with nature and reflects the regenerative values I hold dear. What started out as an ecoretreat center has now evolved into a bioregional hub, a protopia in its own right.

But what exactly is a bioregional hub?

Think of it as a living, breathing center of regenerative activity that's deeply rooted in the natural characteristics, ecology, and culture of a specific place. A bioregional hub works with—not against—the rhythms of the land, water, flora, fauna, and people. It's not just about sustainable living; it's about weaving together social, ecological, and economic health in one place-based tapestry.

At its core, a bioregional hub is a localized ecosystem of solutions. It's a place that brings together regenerative agriculture, ecosystem restoration, community education, ecological design, and economic resilience—tailored to the specific bioregion it serves. Instead of applying generic sustainability frameworks, these hubs respond directly to the climate, challenges, and opportunities of the bioregion they're embedded in. They are nodes of healing, innovation, and cultural revival that help their regions—and the people within them—regenerate.

This is what ChoZen has evolved to become. Located in the heart of Florida's Indian River Lagoon bioregion, ChoZen is so much more than a retreat center; it's a regenerative ecosystem prototype. It sits on over forty acres with protected wetlands and abuts the St. Sebastian River and a 22,000-acre nature preserve home to twenty-six endangered animals. It also sits adjacent to one of the most important watersheds in Florida, the headwaters to the Everglades and St. Johns River, providing water to more than eleven million Floridians. This watershed is under serious threat of human development. As such, it carries a responsibility to protect and regenerate the land, while serving as a learning ground for others to do the same in their own regions.

Through deep ecological stewardship, community gatherings, educational programming, and ecospiritual retreats, ChoZen embodies the ethos of regenerative living. We're actively restoring wetland habitats, regenerating the soil through permaculture practices, and building an intentional community with nature at its core. But perhaps most importantly, ChoZen has become a place where people reconnect to nature and place itself. At ChoZen we have three simple mantras: "Nature is medicine," "Food is medicine," and "Community is medicine."

In a time when so many are uprooted—geographically, emotionally, spiritually—bioregional hubs like ChoZen remind us that healing the planet starts with falling in love with a place and taking responsibility for its care.

The long-term vision for ChoZen is to inspire the development of similar bioregional hubs across the globe, each one honoring the unique ecological fingerprint of its land while contributing to a global web of regeneration. It's not a franchise; it's a framework. And it's one we hope will ripple outward, adapting to the soils, souls, and stories of every corner of Earth.

While ChoZen continues to grow as a bioregional hub, its evolution has been shaped not only by the land but by the people—and the transitions we've navigated together as a community. Like the ecosystems we care for, our human ecosystem has gone through seasons of change, loss, and renewal.

One of the most profound of these turning points for the land came with the passing of my grandmother in 2012, just a few months after she had married me and my wife on November 11, 2011, a celebration not to be forgotten. The ashram went through a challenging period of mourning, and ultimately the community decided they needed to sell some of their land. I felt a deep calling to protect it and keep it in the family, so to speak. That's how ChoZen was born.

Regenerative placemaking is about embracing our differences while working toward the collective protopias we all crave: beautiful, safe places filled with nature, culture, and community. They're not perfect, but they're long-lasting and ever-improving, and spiraling upward. Not all of us can live in temples or in the Amazon, but we can strive to build amazing, sustainable cities and places designed for people from all walks of life to live as one community.

This idea is explored in depth in my recent episode on Gaia's documentary series, *The Road to Utopia: Regenerative Communities*, which takes viewers on a journey across the globe to discover innovative, regenerative models of living. The film highlights ecovillages, self-sufficient communities, and urban projects that integrate sustainability with cultural preservation, demonstrating how we can cocreate thriving environments that honor both people and the planet. Through real-world examples, it illustrates that utopia is not a fixed destination but an ongoing process—one where we learn, adapt, and build together for a better future.

In the documentary, I also delve into my personal journey, sharing a deeper look at life in the ashram where I was raised. Through real footage, I explore the teachings, communal living, and spiritual foundation that shaped my worldview and ultimately led me to embrace regenerative placemaking. By weaving together global insights with my lived experiences, *The Road to Utopia* offers a compelling vision of what's possible when we design places that nurture both human potential and the natural world.

A collective protopia, then, would look and feel a little bit like this: You feel at home in the place where you live. Your community isn't just your neighborhood but a thriving ecosystem of culture and connection. Your environment is one where the buzz of a colorful metropolis harmonizes with the tranquility of nature. The streets are lined with trees, whose lush canopies provide shade and fresh air, while vibrant gardens and parks offer spaces for reading and relaxation, as well as slides, swings, and climbing frames for your children to play together on. Public squares and markets teem with life as local artisans, farmers, and creators share their crafts and produce.

The museums, art exhibitions, and outdoor theaters are brimming with local art and opportunities for creative collaboration. These cultural hubs showcase the talents and stories of the community, offering a dynamic mix of visual art, performances, and interactive installations that inspire and engage visitors. Innovative coworking and coliving environments designed for collaboration buzz on nearly every street corner. These spaces provide flexible and inspiring settings for professionals, artists, and entrepreneurs to work, live, and create together, blending the boundaries between personal and professional life.

The architecture and buildings are designed with regeneration in mind, featuring green roofs, solar panels, and rainwater harvesting systems, as well as dedicated spaces to encourage residents to grow their own food. Public transportation is efficient, clean, and accessible, reducing the need for cars and decreasing pollution. Biking and walking paths weave through the place, connecting people to their destinations while encouraging a healthy, active lifestyle.

Educational institutions are integrated into the urban fabric, offering lifelong learning opportunities and celebrating the diverse heritage of the city's residents. Community centers provide spaces for social gatherings, workshops, celebrations, and collaborative projects, strengthening social ties and a sense of togetherness.

The local economy thrives on innovation, with local businesses and circular economies standing proudly as the lifeblood of the community. All businesses, no matter how large or small, prioritize ethical practices, cultural preservation, and environmental stewardship. Job opportunities are plentiful and diverse, catering to a wide range of skills and passions, ensuring economic stability and growth for all. Every resident and every visitor feels a deep connection to their surroundings, inspired to work together to care for their environment and each other. This is a place where everyone has a voice, and everyone belongs.

Now, ask yourself: *Do you like this idea?* I'd hazard a guess that your answer is yes.

If you're one of the 6.6 billion people who will end up living in a city by 2050,[8] it probably didn't sound just okay; it sounded ideal. And this is not just a fantasy. This is not a dream. This is a regenerative placemaking plan for protopia, and we're witnessing it unfold today.

For the past six years, Future of Cities® has been rapidly prototyping and testing many of these ideas in real-life demonstration projects. Each project is like a living organism that thrives or struggles based on the health of the surrounding ecosystems—places, people, and natural resources all play a role. Our work has become a living laboratory for cocreating protopias. By rewilding the DNA of urban design—prioritizing people over machines and collective growth over individual gain—we are making this protopia a reality bit by bit, place by place.

The more people who believe in regenerative placemaking fight for it and invest in it, the closer we'll get to a future that's bright and secure for everyone.

I trust in this vision completely, as does a growing community of regenerative placemakers and changemakers around the world. Our mission isn't just to prevent our modern world from becoming a flooded Atlantis but to develop a protopia unlike anything seen before in human history.

CHAPTER 2

Why Three Is a Magic Number in Regenerative Placemaking

The Tao gives birth to One.
One gives birth to Two.
Two gives birth to Three.
Three gives birth to all things.

—LAO TZU

Numerology may be a concept that tends to be reserved for spiritually inclined people (and Californians), but when it comes to regenerative placemaking, three really is a magic number. The following rules of three are concepts that will serve you well during your regenerative placemaking journey, and I invite you to combine their guidance as your North Star.

Think Three Generations Past, Three Generations Ahead

From ancient African tribes and Aztec kingdoms to Aboriginal Australians and Māori people, thinking of generations past and generations ahead when making any crucial decisions is rooted in various Indigenous wisdom traditions worldwide.

It's also at the very heart of modern regenerative placemaking. Fundamentally, it's about asking ourselves, *Did this serve generations before, and will it serve the generations that come after us?* and "Are we honoring the past, grounding in the present, and paving a way for a regenerative future?"

Whether we're deciding to build new infrastructure or renovate an old, abandoned space with potential, whether we're choosing cheap, plastic material or higher-quality hemp or recycled steel, we should consider historical examples and future impacts. Will future generations thank us for these decisions, or will they suffer from their consequences?

Beyond what we physically build, this rule can also be applied to creating policies that address climate change, resource management, and social equity with a long-term perspective. It could be applied to any decision we make, for that matter—every day, no matter how big or small. Now that's some conscious forward-thinking.

You can take your forward-thinking as far as you like. Some Indigenous cultures, such as the Haudenosaunee Confederacy, the original people of the northeastern woodlands of Turtle Island (North America), encourage thinking seven generations ahead, a concept embodied officially as the Seven Generations community-based environmental planning tool.[1] No matter how many generations we're including in our decision process or what magical number we're focusing on, this approach of deep reflection and long-term strategic planning is as timeless as it comes.

In practice, this regenerative placemaking rule encourages modern developers to look back three generations to honor and learn from the knowledge, experiences, and lessons, as well as the failures, of our ancestors before making any major development decisions. Through reflecting on the wisdom and ignorance of the past three generations, regenerative placemakers can gain insights into sustainable practices, cultural values, and strategies that have stood the test of time and that can be reused or avoided now.

It also allows us to understand our collective progress, like the huge advancements in healthcare, sanitation, and social security (in primarily privileged countries), and what mistakes we have made as a collective, such as the exploitation of the natural world and Indigenous communities, our contribution to climate change, the breakdown of culture, and the prioritization of individual gain and capitalism, often over the collective good.

Regenerative developers must then look toward the future, considering the long-term impacts on the next three (or even seven!) generations and how we can best serve our future descendants. This foresight is the essence of being a great ancestor. Reflecting on the impact of our actions encourages decisions that prioritize sustainability, environmental stewardship, and overall human and planetary well-being. As you now know, many Indigenous cultures have long embraced this wise approach and, paradoxically, left us, who live in the shiny, new, modern "first world," in the dust. For example, most Indigenous tribes tend to implement sustainable land and resource management practices infinitely better than most governments do. They understand that the health of the environment directly affects the health of present and future generations. They inspire us not only ecologically but socially too. Many Indigenous social structures emphasize collective responsibility, steadfast mutual support, and deep respect for cultural traditions. These elements create resilient societies that are well equipped to handle challenges and adapt to change.

As regenerative pioneers leading a revolution in conscious development, we must reclaim this wisdom. The Ancient Futures movement provides a guiding framework, focusing on integrating traditional wisdom and sustainable practices from ancient cultures with modern approaches to create a more harmonious, resilient future. Popularized by Helena Norberg-Hodge in her 1991 book *Ancient Futures: Learning from Ladakh*, this movement advocates for a return to local, community-based practices that emphasize ecological balance, social cohesion, and a slower pace of life. It critiques the impacts of globalization, industrialization, and consumerism on both the environment and human well-being—lessons I am committed to embedding in my work.[2] Our intention is to revive ancient wisdom while leveraging modern discoveries to their best advantage, ensuring that we leave a legacy of protection and care for life on Earth.

It's curious that pretty much the same regenerative wisdom has been passed down from so many native cultures from every corner of the planet over thousands of years. This widespread, shared understanding suggests a profound, inherent truth: We are meant to care for one another as one interconnected family. This collective wisdom speaks to a universal responsibility that transcends geography, one that urges us to protect Earth and nurture our communities.

While it's true that most of us haven't been exposed to this philosophy, when we embrace regenerative placemaking, we consciously choose to reconnect with this ancient wisdom. In doing so, we get the opportunity to rediscover the essence of who we truly are.

The Three Cos: Cocreation, Codesign, and Coevolution

Imagine you're regeneratively designing a renovation and improvement project for a neighborhood. You've been placed in charge of the entire process. But are you alone? Of course not. It's not just you putting in the effort. You're carrying out your project alongside hundreds of other developers, community leaders, urban planners, investors, residents, artists, architects, and other stakeholders. Yet, you are not regenerative placemaking for yourselves as individuals. Rather, you're working on the neighborhood on behalf of the *whole* community that lives there, including the animals and plant life that call it home. This joint vision relies on three integral concepts of regenerative placemaking: cocreation, codesign, and coevolution.

Just because you're nominally in charge of the project doesn't mean you're above anyone else or that your opinion is superior. You're all in it together as equals. Working together, you can come together as collaborators by putting down your pens and protractors so that you can begin deeply listening to the people of the place you're developing and putting their needs first.

The Beauty of Cocreation

This brings us to the first of the cos: cocreation. Think of cocreation as inviting everyone to the table from the very beginning. Instead of deciding what's best for the community on your own, you engage with the community and all the other stakeholders right from the start. By involving the community in this way, you ensure that the plans you develop truly reflect what the people need and value.

For instance, when planning a new park, community members might express a desire for a playground for their children, a community garden, and an open space for events and farmers' markets with local produce. Their input helps create a park that serves everyone. That's very different from a creative placemaker or group of developers that *assume* what the "perfect" park would look like. As a regenerative placemaker, I've learned that if you *assume*, it makes an *ass* of *u* and *me*!

This idea of not making assumptions is a core teaching from *The Four Agreements* by Don Miguel Ruiz, an incredible read, which emphasizes the importance of not projecting our interpretations onto others.[3] By listening and understanding the community's real needs and desires, we avoid misunderstandings that can result in spaces that don't resonate with the people they're intended for.

Perhaps one of the reasons why my most recent regenerative placemaking project in Jacksonville (PHX-JAX) was so widely embraced by its diverse communities and the city at large was that my team and I stopped and listened to the community first. This led to the community codesigning their best outcomes, and, therefore, it saw a very successful launch.

Cast your imagination to Jacksonville, Florida. It's the largest city by land mass in the contiguous United States—incredibly spread out. This vast city contains stark economic and social contrasts, with a nearly twenty-year life expectancy gap between some Black and white neighborhoods. The urban core has been severely underinvested in, leaving much of its potential untapped. Within Jacksonville is an area called Springfield, and within Springfield, you'll find the most loved and celebrated spot: the Phoenix Arts and Innovation District. Alongside the community and the original artistic visionary of the district, Christy Frazer, we cocreated our latest regenerative placemaking project here.

My team and I at Future of Cities® knew how fundamental the community was in creating a regenerative city that brought Jacksonville back from the ashes. You see, just over a hundred years ago, Jacksonville was literally under a blanket of them. The Great Jacksonville Fire of 1901 was the largest metropolitan fire in the American South. It began on the morning of May 3, when a kitchen fire sparked flames that spread to piles of drying Spanish moss at the Cleveland Fiber Factory—a facility that used the moss as stuffing for mattresses.

Located at Davis and Beaver Streets, the flames quickly spread to most of the downtown area, with smoke visible as far north as Savannah, Georgia. By the time the fire was brought under control, at 8:30 p.m., it had destroyed 2,368 buildings, left 10,000 people homeless, and resulted in seven fatalities.[4]

The name of the regeneration project, lovingly known as PHX-JAX (pronounced *Phoenix-Jax*), is more than just a name. It's a symbol of rebirth. It's a beacon of hope that Jacksonville will rise again, like the mythic firebird. We couldn't have done any of this regenerative placemaking magic without the cocreation that was driven by the community. Throughout the Phoenix Arts and Innovation District project, the team didn't listen only to their own ideas surrounding the built environment, like the upcycling and renovation of existing burnt architecture, but we also listened to the holistic needs of the community and the individual members within it. No voice was dismissed.

From food desert to food forest. From abandoned warehouses to a vibrant, walkable district. The Phoenix Arts & Innovation District in Jacksonville, FL, is a regenerative placemaking project by Future of Cities®—transforming forgotten spaces into thriving hubs of creativity, community, and affordable living. Conceptual rendering by Future of Cities®

Many in Jacksonville have shared their desire for beauty, connection, collaboration, and culture, and we are codesigning our regenerative area of the city with these aspirations at its heart. We are actively nurturing and calling upon an extended community of artists, cultural instigators, and changemakers to uplift the community further and strengthen connections through a shared vision. On October 30, 2024, we celebrated the opening of our first adaptive reuse project—The Emerald Station—a 100-year-old, 18,000-square-foot warehouse located right where the Emerald Trail runs. Jacksonville is on its way to becoming home to one of the most tightly knit, vibrant, creative, fun, and peaceful communities in Florida. And we're getting there by committing ourselves to listening.

The Emerald Station at the Phoenix Arts & Innovation District in Jacksonville, Florida, 2024–2025. Mural by Barbara Hionides

The Art of Codesign

Codesign takes collaboration a step further by involving community members directly in the design process alongside your team. Instead of just asking for their opinions, you work side by side with locals to sketch out plans and develop the actual designs for projects.

Imagine you're designing a combination business center–community center–coworking hub, for example (which Jacksonville community members asked for). First off, you would run a collaborative session where the community members, planners, and designers come together to create detailed plans. During these sessions, everyone sits together, shares ideas, and brainstorms solutions. You should never underestimate the role of community members here, even if they have no background in design at all. For example, if you're designing the community hub, the local residents would contribute their thoughts on everything from the building's layout, the color schemes, and the amenities to the types of activities and services the center should offer.

This is exactly what my team and I did in Jacksonville when we codesigned The Emerald Station, a truly beautiful, inspiring, and collaborative hub, event space, and coworking hotspot. This unique hub is also a 500-person event space packed with potential. The character and history of the building are beautifully preserved, with exposed wooden beams and original brick details, and the walls host some of the most beautiful graffiti in the city. It's modern yet timeless, and it serves as a meeting spot for the entire community to collaborate and to host grassroots events and business conferences with a unique edge. The Emerald Station now forms part of the official Emerald Trail Greenway, and Jacksonville was recently awarded the biggest grant ever given to the city to support our regenerative efforts. (This grant was later rescinded by the Trump administration, but our efforts continue with other support.) The official press conference was held at The Emerald Station, much to the delight of the entire community.

For context, the Emerald Trail Greenway is an amazing initiative designed to connect Jacksonville's neighborhoods, parks, and commercial hubs through a thirty-mile network of green spaces and beautiful pedestrian and bike paths throughout Jacksonville's urban core. The Emerald Trail enriches the lives of the community members through promoting health, fitness, mobility, cultural exploration, and environmental harmony. Once it is fully completed, the trail will serve as a model for urban greenway projects across the country. The Emerald Trail is more than just a series of paths; rather, it represents a bold vision of what incredible, magical heights we can make when we work together and prioritize physical, material, and spiritual connection.

The Power of Coevolution

Coevolution recognizes that a city is not a static entity but a living, breathing whole system that can evolve or devolve, flourish or flounder. Coevolution is about seeing seemingly distinct entities like communities, neighborhoods, green spaces, and architecture as being each a part of a larger whole that can transform for the better together.

Coevolution is about adapting and acknowledging the changing needs, interests, and conditions of places and their communities, also known as *nested wholes*. Nested wholes are dependent on one another, even if they're separate at first glance. Not to be confused with *holes* and actual nests, when I refer to the concept in terms of regenerative placemaking, I'm not talking about homes for our feathered friends. I'm talking about those microsystems within a larger system. For example, within a forest, there are trees. Each tree is a whole and living being. These trees are nested within another whole: a group of connected trees communicating with one another through root-like mycelium (fungus) structures. The birds who live in the trees are also individual wholes. And those wholes are nested within the bigger whole, which is the forest itself.

Imagine your community has asked for an urban garden to grow their own food. This garden might start with just a few vegetable beds, providing healthy, fresh produce for the local residents, who have agreed to tend to that garden as a community. The garden and community are interconnected in this way: The garden provides food, and the local community cares for it. They need each other to thrive and evolve. As the seasons change and as the community's needs and interests evolve, the garden will evolve with them. Perhaps it begins with just including vegetable patches, fruit trees, and local perennial plants that will eventually offer year-round harvests. As more people get involved, educational programs about sustainable gardening and composting could be introduced, say, biweekly in the garden, helping the residents learn how to grow their own food and reduce waste. Over time, the garden is filled with even more veggies and plants, improving the soil's health, strengthening social connections, and providing delicious, fresh, organic food for the entire community.

Eventually, more innovative and sustainable practices are embraced by the community, such as rainwater harvesting systems and renewable energy sources for twinkling garden lights for midnight strolls. The community might also expand to feature community art installations, interactive workshops, and space for local ecofriendly events, reflecting the growing and shifting interests of the ever-more ecologically connected ecocommunity. This kind of dynamic evolution ensures that the urban garden remains relevant and useful to the people who tend to it. Its evolution is inevitable because the community is evolving along with it. They know that when the garden thrives, everyone thrives. As the community flourishes, so does the garden, and vice versa.

This is a parallel that the Regenesis Group uses frequently. The Regenesis Group and the amazing teachers within it are pioneers in permaculture, organizational evolution, and ecological urban design, as well as being a school of education who have a framework course of classes for regenerative development. They teach regenerative soil practices in food systems as an example of how nested wholes (remember, stand-alone but interconnected systems like a garden and a community) can both benefit from each other and evolve together. The Regenesis Group is passionate about this truth: Sacrificing one nested whole for the short-term gain of another is no longer a viable or sustainable solution. Their holistic approach to coevolutionary development is exemplary, and I highly recommend you check out their course, the Regenerative Practitioner Series,[5] if you're interested in finding out more.

Of course, the concept of interconnected systems and nested wholes goes beyond gardens and green spaces. The concept of coevolution is also about lifting up every individual within a community. Imagine a set of Russian dolls. The largest doll represents a country, and the one within it, a city. Inside that is a smaller doll representing a neighborhood. Inside that is an even smaller doll representing a community. Inside that is a family. And inside that is an individual.

When an individual community member thrives—say, opens their own successful bakery–café in Phoenix, Jacksonville—their family's lives change for the better. But then, over time, the entire community and the entire city and the entire country are enriched as a result. Everyone has access to fresh, homemade bread and baked goodies, as well as a social spot to meet each other and catch up over coffee. The bakery–café then gets its regulars, among them, formerly isolated elderly members of the community who now go there every morning as part of their routine. Ten years down the road, that bakery becomes a much-loved local chain, selling delicious pastries all over the city and, eventually, all over the state and all over the country, which impacts even more people. See where I'm going with this?

Through regenerative placemaking, I've seen communities go from resenting each other's successes to celebrating them. They've moved beyond the crab bucket mentality, where individuals hinder each other's progress. Just as crabs in a bucket will pull any escaping crab back down to the depths (this is actually true), downtrodden people with this mindset resist the success of others, preferring to keep everyone confined to the same struggling environment. When we embrace coevolution, on the other hand, we take on a mindset of mutual support and collaboration, focusing on lifting each other up.

Coevolution, therefore, is about creating environments and communities that are not only resilient but also capable of growing and transforming in harmony with their own social systems and ecosystems, which are fundamentally interconnected.

The Three Pillars of Regenerative Placemaking

The third and final rule of three is the cornerstone of any successful regenerative placemaking project. These pillars have become the bedrock upon which my team at Future of Cities® and I are building regenerative cities.

The pillars are community, nature, and culture. Each of these pillars plays a crucial role in creating environments where life can not only survive but truly thrive, both now and for generations to come. They're deeply connected to the earlier principles of generational wisdom and the three cos you just read about.

The three pillars take this particular order for good reason. We start with community because the community *is*, in a huge way, the city, neighborhood, and place. A truly resilient community is one where no one goes hungry, no one is left alone, and no one is left behind. The resilience of a place comes from authentic connections among people. When a community is genuinely supportive and caring, it ensures that everyone's needs are met and that every individual is included and valued. Building a sense of community and understanding the place and its people are crucial before starting any regenerative development. This means taking the time to truly listen to and learn from the area, which is, of course, personified and brought to life by its people. Then, through the community, the designers are exposed to their ecosystems, to their relationship to nature and their surroundings, as well as to their unique cultures. Only with this deep understanding can we implement regenerative projects that make a meaningful and lasting impact.

Dr. Dominique Hes, of the board of directors at Future of Cities® and an adjunct research fellow at the University of Melbourne and Griffith University, describes the effects of the three pillars like this:

> "Ecocities include all aspects of the city, [including] its urban structure and design, its transport systems, as well as its energy, water, waste, money, data, ideas, history, food sources, the larger supporting ecological systems, and its socioeconomic potential. But it's the people who live in the city that are critical to its success and, in fact, its greatest resource. If you have people who love and care for their place, it will foster custodianship, engagement, and all sorts of innovative ways to make a city even greater."[6]

The community, the nature that surrounds it, and the culture that permeates it are intrinsically interconnected and inseparable elements of regenerative placemaking. Each person is nested within the community, the community within the city, and the city within its larger ecosystem and watershed—each whole adding value to and receiving value from the others. If any one of these pillars is left out, crises can ensue. At Future of Cities®, we recognize the vital importance of integrating these pillars into every single project we take on.

In November 2023, we cohosted the Utopian Hours summit, the first US edition of this renowned city-making event, in partnership with Stratosferica. The summit brought together academics, architects, community leaders, and urban practitioners from all over the world to discuss the critical challenges and opportunities facing cities and their people today. Our goals were to highlight the best regenerative practices, share valuable insights from academic research and practical projects, and facilitate new connections among urban development professionals and interested community members.

One of our notable panel speakers was François Alexandre, of KLOTA, which stands for *keeping love on the agenda*. It is an initiative in Little Haiti, Miami, aimed at uplifting communities through various cultural activities. That checks community and culture as covered. But what about nature?

That's where Scott Francisco, founder of the Cities4Forest project, came in. Scott spoke about the fact that by supporting Indigenous communities in their forest-regenerating timber activities, selling low-carbon construction materials to urban markets, and promoting sustainable city architecture, you benefit everyone involved, economically, socially, and ecologically.

Inspired by the innovative process, Haitian American community leader François Alexandre emphasized that Scott was investing in communities, livelihoods, and biodiversity, creating value for the market, local businesses, families, and the Earth itself. Imagine if we could replicate this in every neighborhood, the lasting and resilient economic growth we could enjoy.[7]

The collaboration between François and Scott at the summit was a perfect example of how tightly each of those seemingly different pillars of regenerative placemaking—community, nature, and culture—are woven together. Both of their projects also represent a truth I mentioned earlier: You can create a lot of financial abundance for yourself as an investor and for all the communities involved when you do the right thing and make a commitment to the three pillars.

Regenerative placemaking is a science and a deeply spiritual philosophy, guided by both our hearts and our heads. This way, we have the opportunity to help regenerate life through a bulletproof strategy that embraces emotional intelligence and care for all life on Earth.

CHAPTER 3

Building Bonds Beyond Bricks

Alone we can do so little.
Together we can do so much.[1]

—HELEN KELLER

A place is nothing without its people. In regenerative placemaking, we don't just build buildings; we build community first, which informs what type of neighborhood is needed. A genius team of developers and urban planners could create the most aesthetically beautiful city in the world, but it doesn't mean that city would be a success. After all, beauty is in the eye of the beholder, and redesigning a place without being informed by the existing community could be a disaster waiting to happen.

If you've ever thought that a dedicated placemaking team could create a regenerative place with bricks and mortar and that the perfect people would seamlessly flock there, thank them for it, enjoy it, live in it, and spend their money in it, you're mistaken. I can assure you it seldom turns out this way. The people are the soul of the city, and from them, the built environment springs forth.

Have you ever seen or read the story of Pinocchio? It's a fairytale about a lonely Italian carpenter who turns a block of wood into a puppet to keep him company. This talented man (that's our metaphorical developer—the carpenter, not the puppet) has the bright idea to carve out, chip by chip, a wooden puppet in the shape of a boy. He proceeds to make a wish on a star that he may become a father to the puppet and then expects it to miraculously render itself alive during the night due to the sprinkling of magical fairy dust.[2]

Of course that works in a Disney movie, but it could never happen in the real world. A real man who expects a block of wood to come alive would be delusional, but you get the point. Just like wood generally lacks the ability to transform itself into a living, breathing child, a built environment lacks the ability to transform itself into a real, living neighborhood or city. You have to start with people. The spirit of the people brings a place to life. First comes the lifeforce, then comes the structure.

This principle is especially vital in regenerative placemaking. As Ethan Kent, founder of PlacemakingX and adviser to Future of Cities®, aptly states, "You have to begin by asking what people want to do in a space and then ask how you can support that with design. The community is the expert."[3] Community involvement is not optional; it is integral to the very essence of regenerative placemaking.

This approach, rooted in community engagement, mirrors the principles found in blue zones—regions identified by experts like Dan Buettner where people supposedly live significantly longer and healthier lives. Dan has a documentary on Netflix called *Live to 100: Secrets of the Blue Zones*, which is amazing, and if you haven't seen it, you should. The documentary, as well as his entire book on blue zones, reveals that the secret to longevity may not just be diet or exercise but the strength of our social connections, bonds, and communication.[4] Dan outlines three key elements of connecting that we should be very aware of as regenerative placemakers: a sense of belonging, putting loved ones first, and surrounding oneself with the right community. These principles align closely with regenerative placemaking, which stands for strong, supportive communities as the foundation for spaces that nurture holistic well-being. In fact, on Dan's last visit to our ecoretreat center and bioregional hub in Florida, ChoZen, he declared it to be the ideal blueprint for a blue zone. This was much to our delight, and is thanks to not only the restorative, biodiverse natural setting, but also our focus on building a sense of community between the guests and the locals. Both the blue zone approach and our regenerative placemaking blueprint confirm that putting human connection at the forefront of urban, suburban, and natural design projects makes for the strongest communities of all.

But why else should we start by connecting with and listening to the community as regenerative placemakers? As well as it being the most respectful, ethical, and conscious thing to do, it's highly logical and strategic. Local communities possess knowledge of the place that can't be topped by anyone or anything else. The voices of a community leave theory and expectations in the dust and offer tailored, completely accurate advice and requests for developers. They possess experiential historical information of that place and have personal experiences from recent developments. Only they possess a nuanced understanding of which policies have succeeded and which have epically failed and resulted in suffering, offering critical lessons learned from past experiences that we can use *now*.

This knowledge allows regenerative placemakers to sidestep potential pitfalls and avoid wasting precious time, money, and resources. Additionally, the residents have the insights needed to resolve or prevent social unrest and conflict—issues you definitely want to avoid having on your conscience. They know how to enhance and optimize both individual and collective contributions too, facilitating the development of initiatives that are not only relevant but also highly adaptive and enduring. Engaging with the community ensures that what we build is deeply rooted in local context, constantly evolving, and resilient over time.

Do not make the mistake of assuming that professional developers always know what's best for a city. While developers bring expertise and the ability to make projects happen, it's the local community that holds the keys to meaningful design. Tyson Yunkaporta, author of *Sand Talk: How Indigenous Thinking Can Save the World*, emphasizes key aspects of community engagement—respect, connect, reflect, direct—that are essential to truly understanding and integrating the community's vision into any development.[5] In short, engage deeply and collaborate actively with the local community at every stage of a development project, because your success as regenerative placemakers hinges on it.

For those of you reading this who are part of a community undergoing regenerative placemaking, I urge you to become as involved as possible. Your voice is not just important; it is essential. You possess the lived experiences and local knowledge that the developers, urban planners, architects, investors, and even individual community leaders can't hope to understand. Your input helps guide the project to avoid past mistakes and create solutions that truly benefit the community. Active participation means attending meetings, voicing concerns, offering suggestions, and staying informed about the project's progress. By doing this, you help shape a neighborhood that reflects your collective vision, preserves your cultural heritage, and nurtures a sense of belonging for everyone involved.

Communication Blockers (and How to Unblock Them)

Even though community involvement is essential to regenerative placemaking, bringing together developers, community leaders, project managers, government officials, and other stakeholders with the local residents can be challenging. Getting them to agree and effectively collaborate? Even more so. Let's dive into why this is so difficult and how you can achieve it.

Clashing Priorities and Goals

Imagine developers juggling deadlines, budgets, and project goals. They're often racing against the clock and might focus more on getting the project done quickly and within budget rather than deeply engaging with the community. These professionals might think, *I don't have time to chit-chat*, and might avoid connecting with the community. On the other hand, most community members are emotionally invested in their neighborhood. They care about preserving local history and culture, improving quality of life, and keeping housing affordable, which can sometimes clash with what developers are trying to achieve. When the professionals don't connect with them, the community members think, *These people don't care about us.*

To address these clashing priorities, regenerative placemaking is best served by slowing down, asking the right questions, and listening. Now you know the value of community participation, make room for it in your plan. To avoid unrest or the community feeling disconnected from you, start with a solid, codesigned, shared vision. Host casual workshops and open forums where everyone can share their goals and find common ground.

Neighborhood codesign workshop 2023 at the Phoenix Arts & Innovation District in Jacksonville, FL; Photographed by Bruno Vitale

Embrace the concept of caring loudly. If there is conflict, anger, or even grief, it often signals deep care. These are the voices that have energy for the issue—people who are passionate—although they may not yet know how to channel that energy constructively. It's worth taking the time to engage with them and bring them along on the journey. Sometimes it's best to meet with them separately to ensure they're truly heard without them dominating the larger community discussions. As Bill Reed, Dr. Dominique Hes, and others who work with community-based approaches like FOC and Regenesis have noted, this investment in relationship building can dramatically reduce the timeline and stress. In fact, developers who approach communities this way often find they can streamline processes without the need for extensive legal or PR resources.

Overcomplicated Language

Developers often speak in technical jargon that sounds like a foreign language to most people: "We're integrating ecological design principles, sociocultural dynamics, and participatory planning methodologies here to create an adaptive, resilient, and regenerative environment that has been shown to facilitate your community's holistic well-being." That isn't going to resonate with your audience. Professionals think, *I'll run our meetings with the community like I'd run them with my design team*, but that doesn't work.

Upon hearing this complex rambling, many members of the community will go quiet at best or will feel stupid, confused, and excluded at worst. They are likely to have taken a dislike to these unempathetic developers from the very beginning, and they'll probably not want to collaborate further down the line without something to bring them on board.

To lower the metaphorical gangplank, keep it simple, and keep it human. The professionals can prioritize universally understandable language and can use visual aids like maps and diagrams to bring ideas to life. They can tell stories from similar projects, painting a picture that the community members can connect with. Encouraging questions ensures that everyone feels heard. Most importantly, the community members should not need to know the technical terms of regenerative placemaking to express what they want! Simple language builds trust, connection, and a willingness to collaborate, ensuring that all voices contribute meaningfully to the project.

Mistrust and Historical Tensions

If past developers have broken promises or ignored the community's input, there can be a lot of mistrust and resentment. The community members might see new projects as more of the same, and they may have heard horror stories about *placetaking*, where teams of rich and privileged investors took areas and gentrified them on purpose for profit. Many developers don't appreciate the gravity of this situation. They may think, *We'll make your neighborhood better; just trust us!* without providing any evidence as to why the community should, in fact, trust them. This mistrust from the very get-go can lead to hostility or apathy, making it hard to get people involved.

To prove their genuine good intentions, developers can be empathetic about the community's past. They can research the area's history and that of the surrounding towns. Has gentrification happened nearby? Has this particular community ever been excluded from development projects before? Reassure the community about what your plans are and how they will involve them every step of the way. Developers should share regular updates, be honest about challenges, and stick to their promises. Over time, this can help rebuild trust. There is no universal solution to this kind of earned mistrust; you have to rebuild the trust through honest communication.

Cultural and Socioeconomic Differences

Professional stakeholders might come from different backgrounds and may not fully understand the local culture and values of the community they're working in. They might think, *Pretty much everyone likes the same things in a city. This city could thrive by being redesigned.* The residents, on the other hand, might see the developers as outsiders who don't respect or understand their local culture or way of life, leading to more resistance.

The solution is to build cultural bridges by engaging local leaders who can act as mediators, facilitating communication and understanding between the professionals and the community members. These leaders, with their deep knowledge of the community's history and values, can help translate and contextualize the needs and concerns of the residents.

This is where the essence of a place comes into play, guiding the development of a unifying concept, or *vocation*, as the brilliant Brazilian architect Jaime Lerner called it.[6] This is a broad, inspiring vision that can serve as a North Star, one that every regenerative activist and community leader can rally behind. The professionals should invest time in learning about the cultural heritage and unique characteristics of the area, attending local events and participating in community activities. Demonstrating genuine respect and interest in the community not only builds trust but also ensures that the development project is culturally sensitive and more likely to be embraced by the residents.

Logistic and Practical Barriers

Coordinating meetings and getting everyone in the same room can be tough! Time constraints and logistic challenges often limit meaningful engagement. The professionals might think, *We'll do one meeting per week at the same time, and all the neighborhood can come and share their ideas! Those who don't attend aren't interested.* But the residents might have work, childcare, or transportation issues that prevent them from attending those meetings. They may want to be involved but have too many other responsibilities.

Flexibility, in this case, is crucial. The professionals can offer virtual meetings, provide multiple time slots, and consider on-site childcare during the events. Offering incentives for participation can also help make it easier for people to get involved. Local gift cards or shout-outs can really encourage more people to join in and show them that their time and input are appreciated. Making sure your approach is easy to access, inclusive, and mindful of everyone's needs helps create a friendly environment where community members feel valued and excited to join in. If in-person meetings aren't going as smoothly as hoped—or even if they are—consider designing engaging surveys and interactive web pages with feedback quizzes to complement your approach. In my experience, they work well.

Not Knowing How Important It Is

As I mentioned earlier, some professionals are new to the ideas of regenerative placemaking and simply don't know how important it is to collaborate with the existing community. They think, *It's our job to design and create a beautiful city. Then, as a consequence, the community will enjoy it.* On the other hand, and in my experience, some residents don't know how much they're needed either and how vital their guidance is when it comes to the regeneration of their neighborhood. They may be unaware of the significant impact their guidance can have on shaping a project that truly reflects their needs and aspirations. They might think, *This doesn't involve me and won't impact my life much anyway. Getting involved in meetings isn't worth my time.*

Make sure the community members know that they are needed for all upcoming decisions, and give them real influence. Emphasize that their neighborhood is about to undergo a significant transformation that will affect them and future generations. Create excitement and awareness through comprehensive marketing campaigns using social media, billboards, and public spaces to encourage widespread community engagement and ensure everyone understands the stakes and opportunities involved.

Breaking down these communication barriers takes effort, empathy, and flexibility from both the developers and the community members. By recognizing and addressing these challenges, we can make sure everyone's voice is heard and build vibrant, resilient neighborhoods together. Remember, the heart and soul of any development are the people living there, so make a huge effort to overcome these blockers and connect.

Over my twenty years in regenerative placemaking, I've witnessed the full spectrum of outcomes from urban development. I've seen harmonious, community-led initiatives like PHX-JAX that truly embody the spirit of regenerative development and that are ongoing to this day. On the flip side, I've also seen areas transform from chaos and abandonment to such wild and uncontrollable success that they ultimately became victims of their own achievements. Developments can flourish, art and culture can thrive, and substantial profits can be made. But without meaningful community involvement, these successes can lead to displacement and controversial results.

The key to preventing this lies in joining forces with—you guessed it—the community, from the very beginning. Engaging with the residents ensures that their voices guide the entire development, preserving the neighborhood's soul and preventing adverse consequences.

The Problem of Placetaking and Gentrification

Let me share a story about a place that became a victim of its own success—a place that could have been a complete, 360-degree-turnaround triumph if only the community had been considered more from the beginning. As you're now aware, placemaking is all about revitalizing places with a strong focus on community involvement. On the flip side, *placetaking* refers to the negative side effects that can occur with some placemaking projects that aren't regeneratively focused. It happens when these projects cater mainly to outside groups as opposed to locals, which can lead to gentrification.

For those of you who aren't familiar with gentrification, it's what happens when a neighborhood goes through big changes because new residents and businesses move in. As the area gets a total makeover with fancy housing and new amenities, property values and rents usually go up, making it harder for long-term, lower-income residents to stay. This can lead to displacement, where the original community gets pushed out over time. Along with shifting the cultural and social vibe of the area, it can also drive out original local businesses and change what made the neighborhood unique in the first place. While gentrification can bring improvements like better infrastructure and reduced crime, it raises crucial issues about fairness and the loss of affordable housing. This highlights the need for development strategies that not only drive economic growth but also protect the community's character and ensure that everyone benefits.

Right now, many places face the trend of placetaking because the first pillar of regenerative placemaking—community—is, mostly by accident, left out.

The Magic City Innovation District in Little Haiti offers a powerful example of the complex challenges of large-scale urban transformation. This story goes back over twenty years, to a time when the three pillars of regenerative placemaking—community, culture, and nature—were still largely undeveloped. I share this not only as a success story but as a lesson in facing and overcoming the struggles and addressing the pitfalls that were encountered. This way, you can learn from the development challenges as well as the achievements.

Case Study: The Magic City Project

In the early 2000s, after studying abroad and starting my nightlife career in Argentina, I moved to Miami Beach to become a nightlife promoter. Those were the golden days of South Beach. After a few years of success in promoting nightlife, I got my real estate license and quickly became the top sales associate at my firm as a realtor to the rich and famous. My career blossomed, so I opened my own real estate brokerage, Metro 1, to focus on and colead the urban renaissance in Miami's urban core.

I quickly grew tired of the superficial evolution of South Beach, and many of the creatives had relocated to the urban core. I followed suit. I was increasingly drawn to deeper and more culturally rich projects over there on the other side of the bridge—an area with potential for meaningful change.

This led me to Little Haiti, a culturally vibrant yet severely underinvested neighborhood that had long been neglected by investors, residents, and the city in general. When I arrived, the area was grappling with some serious economic challenges. Small businesses were struggling, many buildings showed signs of decay, and the crime rates were high—all symptoms of a community lacking investment and infrastructure. Yet Little Haiti was also filled with cultural richness and a colorful Haitian heritage to be proud of.

In 2005, I had a vision to transform Little Haiti into Miami's most sustainable, innovative, and culturally vibrant district. I wanted to create a project that would not only drive innovation but also honor and preserve the rich Haitian culture embedded in the area. So I began assembling land, starting with the Magic City Trailer Park, and solicited investors and new partners. We aimed to amplify the neighborhood's beauty through thoughtful urban design, drawing from its deep historical and cultural roots. We named the project Magic City.

Beyond these cultural elements, our team also sought to make significant commitments to support the community's long-term growth. We committed US$41 million to the Little Haiti Trust, comanaged by the city and community leaders, to fund the initiatives that mattered most to residents. Additionally, we lobbied for a new commuter train stop to be part of the project, which is being planned as we speak, connecting Little Haiti with downtown Miami and enhancing public access across the city. The project was successful in being included in the Opportunity Zones when the legislation was passed. These efforts aimed to bring essential infrastructure, funding, and opportunities directly to the neighborhood, reinforcing Little Haiti's connection to the larger urban landscape and supporting its cultural and economic resilience.

Then there was the *official* name change. At that point, some still referred to the area as Lemon City. The team supported the initiative to formally name the area Little Haiti, giving Haitian residents a place to identify with their history and culture, as opposed to keeping Lemon City, the sorely outdated original name. Before becoming the culturally rich area it is today, Lemon City had ties to an agricultural past, where enslaved people formed part of the workforce in the local lemon groves.[7] This history contributed to the area's eventual shift in identity, paving the way for Little Haiti to emerge as a place of cultural pride. Little Haiti's transformation attracted visitors from all over, and the area improved—became safer—but the small businesses continued to struggle despite the new investment.

Despite our best intentions and incredible concerted outreach to a lot of community members, we had mixed results when it came to local support of the project. With differing visions, special interests, and opinions among the existing residents, it was tricky to navigate the conflicting needs within the community, even though we got approval for the project from the city council. This led to certain challenges. Added to those challenges, we were unable to secure necessary incentives to offer low-income housing in the form of affordable and workforce housing, which is so badly needed in this area. Providing these options could have created greater alignment between the community and the developers, strengthening trust and collaboration in the process.

My connection to Little Haiti began long before my real estate career—working in the nonprofit sector as president of the Non-Violence Project alongside founder Diane Landsberg, where we helped provide conflict resolution education to at-risk youth at schools like Edison Senior High. I've always had and will always have deep respect for this neighborhood, its rich cultural history, and the strong community that existed long before it was officially designated Little Haiti.

Like any evolving place, hope is never lost. Regenerative development is about recognizing potential and coevolution. The Magic City project has brought investment, creativity, infrastructure, and attention to the area—elements that can lay the foundation for a thriving future. No development project develops in exactly the way we envision, but with continued effort and a regenerative approach, Little Haiti can be viewed as a success story that benefits the entire community.

Recently, I met with Joann Milord, the executive director of the Little Haiti Trust. We reminisced about the journey we had embarked on and how the area's renaming and US$6 million initial investment has ultimately benefited the residents of Little Haiti. Through its recent first-time home buyer programs and repair initiatives, real positive change is starting to happen.

This experience has been instrumental in shaping the Future of Cities® framework, which places community as the first pillar. Our lived experience has taught us the importance of inclusive, community-led development rooted in respect for local history and underscored by the need for affordable, regenerative, nature-focused design. Projects like Magic City are a reminder that even with the best intentions, true regeneration requires listening, collaborating, and creating spaces that are shaped with—and not just for—the community. And it does require support and incentives from the public sector.

Placemakers can learn valuable lessons from Little Haiti and gain major insight into how easy it is to generate unjustified unrest and angst alongside success. As my friend and the Haitian American community leader of Little Haiti, François Alexandre, wisely shared at one of our events, "We needed a hand up, not a handout."

To avoid pitfalls like these, Future of Cities®'s approach to regenerative placemaking focuses on integrating the pillar of community, followed by nature, followed by culture, from the very start. That's how you go from creative placemaking (risking placetaking) to *regenerative* placemaking.

Tanya Watts, director of neighborhood affairs for our Phoenix Jacksonville Arts and Innovation District, emphasizes that we want development to happen in partnership with the community.[8] This approach is central to the project. Our goal is to keep strengthening connections among the residents, artists, businesses, and public spaces and to redefine urban spaces to support empowerment, innovation, and the arts. And it all starts with treating the community as the experts and developing based on their ideas, requests, and advice so everyone can thrive.

Withintrification: The Antidote to Gentrification

We call that total community involvement *withintrification*, a unique concept born from regenerative placemaking. It's a rebellion against the prevalent challenge of gentrification and a value that leads to holistic success for everyone involved. It's a concept that focuses on revitalizing neighborhoods from within, ensuring that the area is improved and upgraded without displacing its current residents. Unlike traditional gentrification, which can end up pushing out longtime locals to make way for generally wealthier newcomers, withintrification aims to enhance the quality of life and infrastructure of a place in a way that respects and preserves the existing community's cultural and social fabric. This involves—you guessed it again—engaging closely with local residents and stakeholders throughout the planning and development process to ensure that new projects align with their needs, desires, and values.

By prioritizing the community's input, withintrification and regenerative placemaking improve public spaces, housing, and amenities while maintaining affordability and accessibility for the existing residents. This means that infrastructure upgrades and new developments are designed to benefit those already living in the neighborhood, and it also attracts positive outside attention. This prevents the fragmentation often caused by displacement and totally uplifts and improves neighborhoods in ways that are inclusive and supportive of the community, one that honors its past while looking forward to a sustainable and abundant future.

In other words, withintrification empowers the existing residents to lead the regenerative development process. This highly collaborative approach preserves neighborhood identity and mitigates displacement risks and developer–community tensions too. Our regular community meetings in the Phoenix Arts and Innovation District aim to mitigate gentrification by always prioritizing what's best for the current residents.

Case Study: The Phoenix Is Rising

Before I let you in on the nitty-gritty of how we are helping to nurture and build a great relationship with the people of Springfield, Jacksonville, let me tell you the intriguing story of how it all came about. In the last chapter, you read a little about The Emerald Station, which was birthed from the three cos: cocreation, codesign, and coevolution. I also told you about how, once upon a time, the entire city was on fire, leaving nothing but ashes and crisis.

Crisis, though, can lead to total transformation. In Chinese, the word for "crisis" (危机) is frequently highlighted to illustrate this idea because it combines the characters for "danger" (危) and "opportunity" (机). Deciding to take a chance on Jacksonville and dedicate years of my life to the area posed the danger of failure. No one had ever attempted to introduce full-scale regenerative placemaking to that area before. However, our collective intuition at Future of Cities®

revealed a tremendous opportunity waiting to be embraced. And it all started with a phoenix that landed on my back.

In 2019, I arrived in Jacksonville just after my birthday. A few months earlier, I had been basking in the beauty of Nukutepipi, a small, secluded private island in French Polynesia. It was owned by one of my ex-business partners, Guy Laliberté, the same visionary founder of Cirque du Soleil who was a stakeholder in the Magic City project. He had invited my wife, Ximena, and I to stay to celebrate three major events in our lives: our joint fortieth birthdays, our seventh wedding anniversary, and Ximena going into remission after having breast cancer. The invitation to relax at the island resort was an extraordinary gift, and the experience was nothing short of magical. As a final present to all of the guests he was hosting at the time, Guy offered complimentary tattoos to be drawn by the most talented Polynesian tattoo artists in the country. Neither Ximena nor I had ever seriously considered getting a tattoo. But the ambiance of the island and the conversations we had with the talented tattoo artists inspired us to change our minds. Ultimately, I decided on a phoenix, a powerful symbol of rebirth, which seemed fitting as I was beginning a new phase of my life: I would be entering into the world of regenerative placemaking and leaving my real estate broker days behind.

When I arrived home, I saw my chief investment officer at Future of Cities®, Ralph Davies, pop up on my phone. Ralph had a keen eye for spotting opportunities and was skilled in research and data analysis. He called me to share, very enthusiastically, that Jacksonville had a positive demographic trend, with millennials and young people increasingly moving downtown. Ralph saw huge potential for

Jacksonville to become a hub for artists and innovators. His enthusiasm was infectious, and I soon found myself planning a scouting trip to explore the possibilities. When I arrived in Jacksonville, I followed my instincts.

I noticed that the houses were so different from what you'd see in other parts of Florida. There were traditional porches out front, where a few people would play music, eat together, and connect. They even had a porch festival every year, according to one of the locals. There was a feeling of openness, and unlike a few other places in Florida, the neighbors weren't hidden behind large fences, Astroturf, nonnative plants, and garages. This was traditional presuburban America, like a welcome flash from the past, and I liked it. I later found out that those houses were built in the years following the Great Fire of 1901, when the fire destroyed much of downtown Jacksonville and residents relocated across Hogan's Creek to Springfield. Decades later, the Springfield Preservation and Revitalization (SPAR) Council was formed to protect and restore what had been built, and the area was officially designated the Springfield Historic District.

I kept wandering the streets, and I ended up in the office of SPAR. I was greeted by Kelly Rich, the executive director at the time. Kelly's passion for Jacksonville was evident from the moment we met. She spoke about the district's evolving landscape and the pivotal role of the arts in this transformation. "You have to meet Christy Frazer," she insisted. "She's the heartbeat of the arts movement here. She owns some warehouses and has a bar that's become the hub for a ton of underground events, and she wants to expand the artistic community here."

Bingo.

Intrigued by Kelly's recommendation, I sought out Christy, and it didn't take long for me to realize that her vision aligned perfectly with our mission at Future of Cities®. She was a fascinating woman: an artist as well as an entrepreneur and owner of multiple warehouses in the area. Her grassroots approach to artistic community building and my strategic expertise in regenerative placemaking were a match made in heaven. My gut was telling me to move forward with it and take a chance on Jacksonville. That's when I asked, "What area of Jacksonville are you based in?"

She smiled. "Phoenix."

I suddenly felt the dull burn of the ink from my new tattoo.

She continued, "I've got a feeling that, with your regenerative placemaking experience and my connections with the artistic community, we could help this phoenix rise from the ashes!"

Christy and I had both read the tea leaves, and we went on to invest in her vision. Together with my team at Future of Cities® we ended up buying Christy's collection of three warehouses and embarked on a journey to bring regenerative placemaking to Phoenix, Jacksonville. We were lucky, because thanks to Christy and SPAR, and their former efforts to rally together local artists, we already had a small artistic community to work with. Jacksonville had proven to be an up-and-coming place for talented artists, and we believed that the future of Jacksonville, much like the phoenix, was destined to soar.

But it wouldn't happen overnight. The next step in our journey was to bring about a larger, diverse community and carve out a shared vision. Our goal was to ensure that the revitalization efforts were not just top-down initiatives but genuinely reflective of *all* the residents' needs, not just the artistic ones.

In the beginning, one of the biggest challenges was building trust. Some residents, wary of past promises unfulfilled and the risk of gentrification, were understandably cautious. We addressed these concerns head-on by communicating that the community would be at the heart of the project, especially community leaders, who would be on the frontline of our designs. We established a community advisory board, including representatives from different sectors and backgrounds, to oversee the projects and ensure accountability.

Getting everyone physically together in the first place and finding an accessible spot was also a challenge! That said, we organized a series of community meetings, inviting residents, business owners, artists, and other stakeholders to share their thoughts and ideas. These gatherings were held in various local venues at different times—from Christy's place to community centers and online—ensuring accessibility for everyone. We opened the floor completely to the residents. Their input was invaluable and revealed a deep love for their neighborhood and a collective desire for positive change. Some craved more green spaces and gardens to grow their own food, others emphasized the need for affordable housing, and many called for greater support for local businesses, events, and artists. Together with the community, we created our multilayered mission. We handwrote our Future of Cities® manifesto in one of our meetings like this:

> The mission of Phoenix Arts and Innovation District is to sustainably build equity through community, arts, nature, and culture and provide a global platform for business incubation and innovation in Jacksonville's North Springfield

> neighborhood of Phoenix. The team at Future of Cities® is dedicated to cocreating a community with artists, residents, cultural instigators, and changemakers to provide access to educational opportunities, innovative programming, technology, and community events. We are committed to building a vibrant and diverse neighborhood.

One of our most significant achievements was, of course, The Emerald Station, built on the dreams of the community, and a space for the entire neighborhood to coelevate one another and thrive. But we didn't just build structures; we also facilitated amazing in-person events. A standout moment for the people of Jacksonville was the Phoenix Holiday Art Fair and Makers Market at PHX-JAX, held in December 2023. We hosted the first annual event in a repurposed industrial warehouse, and it was a massive success. Over fifty local businesses, artists, and vendors showcased their work to more than 1,200 attendees. This market didn't just boost local economic activity; it also enriched the community culturally. It was a space where people from various backgrounds came together to celebrate their heritage and a newfound sense of community.

Inspired, we went on to plan something incredibly beautiful and never done before. From September 29 to October 1, we transformed 14th Street into a vibrant open-air gallery with Jacksonville's first 48 Hour Mural Festival. Partnering with the Jacksonville Arts and Music School, we invited local artists to work side by side, accompanied by live music, to paint stunning brand-new murals to celebrate the area's revival.

The results breathed life, color, personality, and legacies along

14th Street. Emily Moody, our director of community engagement at PHX-JAX and Future of Cities®, discussed the placemaking effect: "People don't always just wander into a museum or find access to public and private art galleries. What we achieved was essentially creating an open-air public art museum for residents and visitors alike to enjoy, and feel the cultural and creative nature of our communities. . . . It was a joy to see people coming together. The neighborhood residents and community members at large, driving by, getting out of their cars, walking around, and engaging with the artists. It was powerful and inspiring to not only witness muralists in action but also share a conversation with them and hear the story behind their work of art."[9]

During 2024 alone, we hosted another thirty-six free events and welcomed more than 5,000 people to the district. Through these initiatives, we witnessed firsthand how regenerative placemaking can transform struggling neighborhoods into thriving hubs of culture and creativity. The success of these projects reinforced one crucial lesson: When cocreating with communities, success moves at the speed of trust. The ongoing journey of helping to transform Jacksonville's Phoenix Arts and Innovation District, although we're far from done, is a testament to the power of regenerative placemaking, showing how honoring a community's voice and empowering its residents can lead to incredible results.

Success Moves at the Speed of Trust

As we've explored, the essence of regenerative placemaking is deeply rooted in the community. It's about more than constructing buildings and designing spaces; it's about nurturing the spirits of the people who live there. When working with communities in any developmental project, the key to this transformation is trust. Trust is not just a luxury or an afterthought; it's the bedrock of regenerative placemaking.

This process isn't just about inclusion; it's about empowerment. Trust cultivates a sense of ownership and pride among the residents, which strengthens the social fabric and resilience of any neighborhood. Moving forward with regenerative placemaking projects, all of us would do well to commit to honoring and building this trust. Every decision, every plan, and every action must be made in partnership with the community. As our community strategist at Future of Cities®, Debra Webb, says, "Placemaking has immense potential to create belonging, address challenging social issues, connect people to each other, and strengthen their communities." I believe it's accurate to say that the real strength of our cities lies in the trust and collaboration we build with the existing community of a place, as well as the interconnectedness of people and their environment. To cocreate truly regenerative places, we must first embrace the need to become regenerative ourselves individually, then collectively. This echoes the wisdom of a quote I dearly love, commonly attributed to Rumi: "Yesterday I was clever, so I wanted to change the world. Today I am wise, so I am changing myself."[10]

The transformation of our surroundings begins with our own personal then a shared regenerative mindset. When individuals and communities embrace collective, regenerative identities, this positive change radiates outward.

Effective Community Engagement

Building trust starts with listening to the community. There are three preliminary steps I'd recommend you take. When planning and conducting community meetings, you should consider the following to streamline and optimize your process:

Notification and Representation

How are communities and stakeholders informed about upcoming meetings? Consider whether the communication channels used are the most effective ones for reaching your target audience. Are the notifications being distributed through the mediums that community members typically rely on for news and events?

The location of the meeting is crucial. Is it convenient and accessible for the target demographics? Ensure that the location is central and easy to reach, particularly for diverse neighborhoods, to encourage broad participation.

Think about the diversity and inclusivity of the participants. Does the meeting represent the full spectrum of the community? Efforts should be made to ensure that all relevant groups are represented. To support this process, don't be afraid to get the most passionate people—those with big energy, be it positive or negative—involved. In my experience, once the group begins to effectively work together, this attracts outside attention and the diversity begins to develop naturally.

Meeting Process and Documentation

Once meetings take place, it's important to gather a wide representation of feedback. How will this feedback be documented and communicated to the decision-makers? Clear documentation ensures that the voices of the community are accurately conveyed and considered. When it comes to gathering feedback, don't hesitate to share with the community. Let them know this is what you heard from them, and this is how you're responding.

What activities will be included in the meeting? Will it be an open Q&A session, or will the participants be asked to brainstorm solutions and articulate their needs and aspirations? The structure of the meeting can significantly influence the quality of the feedback received.

Data Collection

Be prepared. Before the meeting, it's important to consider what kind of information and knowledge is necessary to achieve the project's goals. Look at the quantitative data: How will you collect measurable data? Consider using surveys, scaling questions, or multiple-choice formats to gather quantitative insights. But also pay attention to qualitative data: For deeper insights, qualitative data collection methods such as interviews and focus group discussions can be employed. These methods can help capture the nuances of community perspectives and aspirations.

By taking these considerations into account, you're making sure that your community engagement process is thorough, inclusive, and effective in gathering valuable insights that contribute to the success of your project.

CHAPTER 4

Embracing the Green Pulse

Nature is not a place to visit. It is home.[1]

—GARY SNYDER

Nature is more than just a retreat for weekend visits or a pleasant backdrop to our urban lives. Nature is our mother, our home, and the source of all life on Earth. Nature cannot be defined by one color nor can it be constricted by plastic plant pots and fences. It's more than nice to have. It's more than decoration and more than a resource. We need nature around us to survive, and we need it to thrive. But what many people don't realize is the following truth: We *are* nature.

Humans aren't separate from nature, with *us* over here and *it* over there; we are an intrinsic part of nature, a living, breathing expression of it. As the philosopher Alan Watts once said, "You didn't come into this world. You came out of it, like a wave from the ocean. You are not a stranger here."[2]

The elements in our bodies were forged in the stars, and when those stars exploded as supernovae, they scattered building blocks across the universe. Over billions of years, these elements divinely collided to form our planet and, eventually, the diverse dance of life itself. We are a continuation of all forms of life that share this planet. Every living creature, from the grandest mammals to the smallest insects, bears the imprint of these same origins. The birds that soar through the sky, the reptiles that bask in the sun, the amphibians that thrive in the wetlands, and every single plant that unfurls their leaves are all bound by the same heritage: our collective, natural heritage. We are not separate from nature but deeply embedded within it and connected to it. We, as well as every single other manifestation of life, are integral parts of the universe's grand design for this planet. And we didn't come this far to chop it down, pollute it, sever our ties with it, and confine ourselves to varying concrete boxes we call home.

This shared heritage bonds us in a way that cannot be undermined or ignored. We have a deep desire to be close to nature itself and, therefore, to our true nature. Have you ever been on an intense city block with no green spaces and then felt an unwavering craving to go hiking in the mountains after you got home? Perhaps once upon a time, you were working so hard at the office that you forgot to spend time in nature, and, eventually, you experienced a painful withdrawal and an overwhelming urge to connect to it. That's because you're a biophile.

> *Biophilia*: a hypothetical human tendency to interact or be closely associated with other forms of life in nature; a desire or tendency to commune with nature.

Biophilia is built into us. This whole idea of somehow living without it or separately from it, therefore, is impossible. Biophiles also need more than an infrequent dose of nature on special occasions. Biophiles need to live alongside it, as part of it. This is why nature is the second pillar of regenerative placemaking. A chronic dismissal of nature's importance in our cities and neighborhoods won't just make our own lives suboptimal; it will also leave future generations much worse off.

In 2018, Plant the Future transformed an old gas station and sports bar into a lush biophilic design destination in Miami's Little River, including a design shop, cafe, and venue—their first commercial real estate purchase through adaptive reuse.

The Regenesis Group stresses that we must work with nature, playing an active role in protecting, sustaining, and allowing it to evolve alongside us in the places where we live. In this context, sacrificing nature and exploiting its resources with the excuse of benefiting humans just isn't cutting it anymore. When we harm one ecosystem, we harm another, and each is connected and inseparable.[3]

The shift in mindset from individualism to recognizing the interconnectedness of nature and all systems on Earth is often called *ecological thinking*. Carol Sanford, a pioneer in regenerative development in the business world, explains an important difference between environmental and ecological thinking. The term *environment* suggests a context or background where something exists, implying a division between *us* and *not us*. In contrast, *ecology* views everything as part of a dynamic, interconnected whole. This perspective emphasizes how all elements within an ecosystem influence and interact with each other, highlighting the importance of seeing the world as an integrated and interdependent system.[4] This way of thinking encourages us to move beyond seeing ourselves as separate from nature and instead understand our role as a part of a larger, interconnected web of life.

A recent project that truly embodies this connection is *Path of the Panther*, a National Geographic documentary that my wife and I had the privilege of contributing to, alongside executive producer Leonardo DiCaprio. This Emmy-winning film tells the gripping story of the fight to conserve millions of acres within the Florida Wildlife Corridor—an effort that is critical not only for biodiversity but also for the long-term resilience of our planet.

Path of the Panther follows a coalition of wildlife photographers, veterinarians, ranchers, conservationists, and Indigenous leaders as they work together to track and protect the endangered Florida panther. National Geographic photographer and filmmaker Carlton Ward Jr. uses camera traps to capture breathtaking images of these elusive big cats and other marshland creatures, while biologists monitor their movements, veterinarians rehabilitate panthers injured by car strikes, and ranchers safeguard working lands to prevent overdevelopment.

The film highlights a bold vision for the future: creating interconnected wildlife corridors across ranchlands and nature preserves to restore and protect the panther's dwindling habitat. Through stunning cinematography and deeply moving storytelling, *Path of the Panther* makes a powerful case for the role of conservation in regenerative placemaking. For me, it represents what I call *regenerative storytelling*—stories that don't just entertain or inform but actively drive ecological restoration and awareness. This film serves as a compelling reminder that regenerative placemaking is about more than just buildings and urban spaces; it's about protecting the ecosystems and keystone species that sustain life itself.

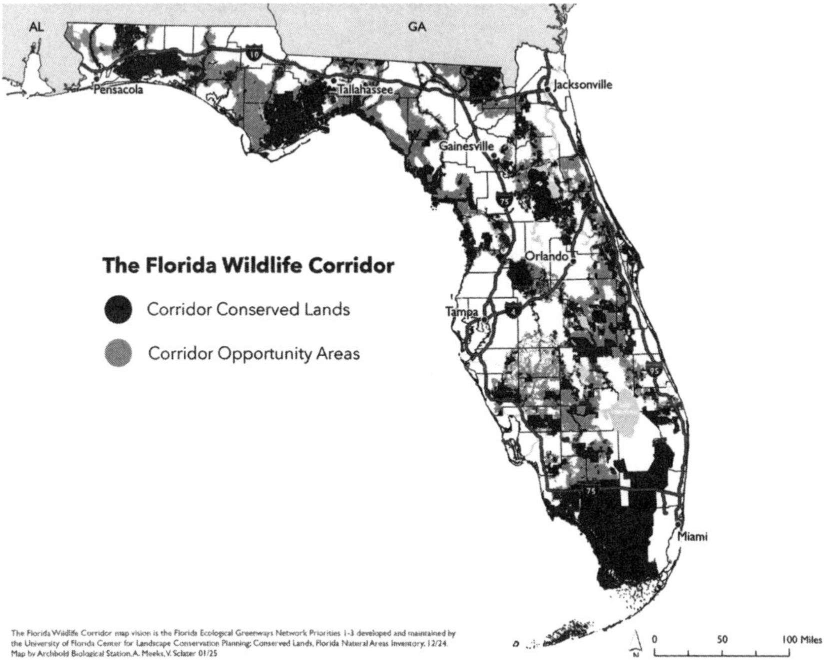

A map of the designated area of the 18-million-acre Florida Wildlife Corridor—a connected corridor of public conservation lands and private working lands throughout the state of Florida; Provided by Florida Wildlife Corridor Foundation

A Florida panther "Babs" and her cubs. The first female panther documented north of the Caloosahatchee River since 1973 and a pioneer representing the expansion of her species' historic range; Photograph: Carlton Ward Jr. / Wildpath and *Path of the Panther*

This idea of human connection to nature goes even deeper when we talk about the overview effect. Astronauts describe this profound shift in consciousness when they see Earth from space—a sudden realization of the planet's fragility and our interconnectedness with it. At our ChoZen Eco-Retreat and Center for Regenerative Living in Florida, we bring that feeling to life through our ChoZen Air seaplane ecoadventures narrated by Carlton Ward Jr., the same NatGeo filmmaker who cocreated *Path of the Panther* with Eric Bendick of Grizzly Creek Films. On our flights, visitors soar over the vast wetlands, tracking the majestic and endangered Florida panther, and dive into the area's natural beauty. It awakens a sense of ecospirituality, reminding them that their relationship with nature is deeply rooted within them, even if it has lain dormant for a while. And it's this kind of awakening that can inspire people to take action in ways that truly regenerate our world.

The bottom line is this: We are meant to live in and around nature. We *are* nature, not apart from it. We are meant to prioritize it, learn from it, evolve with it, and connect to it every day. When we devalue it, there are consequences for us humans as well as for all other life on Earth.

Consequences of Disconnecting from Nature

Trigger warning: What you're about to read is quite sad, yet it's a reality of what nature is going through right now. If we truly are nature at our core, then these are crises we all face. This information is crucial for you, both as a human and as a regenerative placemaker, to deeply understand why nature is the second pillar of regenerative placemaking. So as you read through this, don't lose hope. Regenerative placemaking holds vital keys to start living in a new way. But in order to solve problems, we need to know what the problems are first. Let's explore.

Goodbye, Biodiversity

Since the Industrial Revolution, our disconnection from nature has caused many people to devalue it. Our actions have increasingly scarred and degraded the world's vital ecosystems. Forests, grasslands, wetlands, and other crucial habitats are vanishing at an alarming rate, jeopardizing our own well-being. According to the World Wildlife Fund's *Living Planet Report*,[5] 75% of Earth's land has been significantly altered, our oceans are heavily polluted, and we've lost more than 85% of our wetlands. This relentless destruction has pushed around a million species—including 500,000 animals and plants and 500,000 insects—toward the brink of extinction.

The most pressing driver of biodiversity loss in recent decades has been the conversion of pristine habitats—forests, grasslands, and mangroves—into agricultural lands. Our oceans have suffered from overfishing, a practice that continues to deplete marine life. Since 1970, these trends have been fueled by the doubling of the global population, a fourfold expansion of the global economy, and a tenfold surge in trade.[6]

Apart from this being very sad, the loss of biodiversity for humans is bad news for many reasons. Here are a few:

- Many of the plants that feed us rely on animal pollinators like bees and butterflies. Without these pollinators, crop yields are plummeting, threatening our food security.
- Wetlands and forests filter pollutants from water, ensuring clean drinking water. Their degradation can lead to more polluted water sources (as well as higher treatment costs).
- Human encroachment on natural habitats increases the risk of zoonotic diseases—diseases that jump from animals to humans—by disrupting the balance between wildlife and disease vectors.[7]
- Forests and other ecosystems absorb carbon dioxide, mitigating climate change. Their loss exacerbates global warming, leading to more extreme weather events and rising sea levels.

For example, natural wetlands, like those found at our ChoZen Eco-Retreat and Center for Regenerative Living in Florida,[8] serve as huge carbon sinks, and although they only cover about 1%–4% of Earth's surface, they store approximately 20%–30% of the world's organic carbon.[9]

Hello, Climate Change

That brings us to our next challenge. Another consequence of neglecting our connection with and protection of the natural world is climate change. While many of us are familiar with the causes, let's recap the situation.

Human activities have drastically altered our climate. Rising temperatures are a direct result of increased greenhouse gas concentrations, exasperated by, for example, deforestation. When we clear forests for agriculture, logging, and urban expansion, we not only reduce Earth's ability to absorb carbon dioxide but also release stored carbon back into the atmosphere. Trees and vegetation act as crucial carbon sinks, capturing carbon dioxide and helping to regulate our climate. Their destruction exacerbates global warming. Our energy consumption also plays a critical role. Burning fossil fuels—despite our awareness of its environmental impact—releases large amounts of carbon dioxide, further heating the planet. Poor waste management practices compound the problem further, with methane emissions from landfills adding to the greenhouse effect.

While it's true that Earth has experienced natural warming and cooling cycles over millennia, such as the transitions in and out of ice ages, the current rate of warming is unprecedented. Historically, it took about 5,000 years for the planet to warm between four and seven degrees Celsius after an ice age. In contrast, the past century has seen a temperature increase of approximately 0.7 degrees Celsius, a rate roughly eight times faster than the average post–ice age warming.[10]

These elevated temperatures are causing polar ice caps and glaciers to melt, leading to rising sea levels that flood coastal areas around the world, including places like Florida. But the impact of climate change extends beyond flooding. It is linked to an increase in the frequency and severity of extreme weather events, such as hurricanes, droughts, fires, and other natural disasters. Take the recent Hurricane Helene, for example. It didn't just devastate Florida; it also wreaked havoc in the remote mountains of North Carolina, places you wouldn't expect to be hit so hard. Our partner Carlton Ward Jr., the NatGeo photographer and filmmaker we mentioned earlier, experienced this firsthand. On the very same day that we won an Emmy for *Path of the Panther*, his home in Tampa was flooded. The irony is that the documentary—and Carlton's work in general—is focused on conserving millions of acres of the Florida Wildlife Corridor because these natural ecosystems, you guessed it, protect us from flooding. It was a stark reminder of how urgent our work is and how quickly nature is responding to our actions—or inaction. These extreme conditions contribute to disrupted habitats, altered food chains, and the loss of critical ecological functions that threaten the survival of countless species.

Apart from this being very sad for the animals and plants that cannot survive in a world with such extreme climate change, here's the reason it's scary news for us: We can't either. At best, climate change will turn life on Earth upside down. At worst, however, it could, at some point that's unknown, be the end of us.[11]

Goodbye, Peace of Mind

If widespread destruction and potential extinction aren't enough incentive to make nature a priority while designing our cities, choosing our materials, and carving out our way of life as a collective community, not being connected to nature as part of our lifestyle can also take a toll on many people's mental health. A lack of contact with nature is increasingly recognized as a significant factor contributing to mental health issues such as anxiety and depression. Studies have consistently shown that exposure to natural environments has a positive impact on mental well-being, while its absence can exacerbate symptoms of mental illness and, over time, cause psychological distress.

We were inherently designed to coexist with nature, living as seamlessly as possible within its cycles and rhythms. Our evolutionary path has been intertwined with the natural world, shaping our psychology and what we need in order to feel happiness and peace. Exposure to the sun and moon dictates our circadian rhythm, which contributes to good mental health. Exposure to trees lowers blood pressure and stress. Spending time at the beach or in forests lowers cortisol levels. All of this lowers anxiety and promotes 360-degree well-being. This cannot be attained in most urban environments in large, nonregenerative cities.[12]

Studies by a variety of scholars spanning multiple disciplines have proved that the natural environments we have contact with impact human happiness. Gregory N. Bratman, of Stanford University, investigated the effects of nature exposure on mental health by comparing individuals who walked in natural environments with those who only walked in urban settings (without any parks or greenery). The study showed that participants in the natural environment reported significantly lower levels of rumination (a repetitive thought process linked to depression) and showed decreased neural activity in brain regions associated with depression. This evidence supports the notion that natural environments have restorative effects that mitigate anxiety and improve overall mental health.[13]

At our ChoZen Eco-Retreat and Center for Regenerative Living, we actively demonstrate regenerative practices that improve the mental and physical health of humans, protect biodiversity, build resilience, and empower communities to thrive in balance with nature. As a nonprofit organization, ChoZen Eco-Retreat and Center for Regenerative Living is dedicated to promoting regeneration through nature-based experiences, sustainable agriculture, educational programming, artist residencies, land and wildlife conservation, and retreat scholarships for those who need it most.

By integrating natural spaces into daily activities, ChoZen exemplifies how regenerative placemaking can promote both individual and collective health, aligning with a vision rooted in healing through nature. In this balanced state, we find a deeper connection to ourselves and to our natural world, bringing about a beautiful sense of belonging that modern, artificial environments often fail to provide. Moreover, studies have shown that spending just two hours in nature can lead to significant reductions in stress levels, with lasting effects over the course of *two weeks.*[14]

The bottom line is, in order to be happy, at peace, and mentally healthy, human beings need nature.

Goodbye, Physical Health

But it's not just our minds that benefit from nature. It's also our bodies. Physical and mental health are, of course, linked. But not living in and around nature and natural spaces is increasingly recognized as a trigger of physical health issues for humans too. What's the use of hitting the gym three times per week if we're starved of one of the main sources of physical well-being? Research backs up the fact that living life surrounded by urban sprawl and disconnected from nature can contribute to multiple ailments:

- Obesity, caused by a lack of activity; people who live away from nature are much less likely to exercise, and therefore become unhealthily overweight[15]
- Cardiovascular diseases, caused by the same conditions as obesity[16]
- Vitamin D deficiency due to lack of sunshine (which increases the risk of death by COVID-19[17])
- Poor sleep and lowered immune system functioning[18] due to artificial lights, and being out of sync with nature's cycles
- Reduced respiratory health in children, causing asthma and impaired lung growth[19] due to air pollution

On the topic of pollution, if you watched the recent Olympic Games in Paris, you'll know that even elite athletes aren't immune to its damaging effects. In the summer of 2024, during open water races, Olympic swimmers battled not just the tides but highly contaminated water (which they were assured was "clean and safe enough to swim in"), falling ill from bacteria and viruses lurking beneath the surface.[20] (The Seine has since been officially declared safe to swim, which gives me hope for regeneration at the urban level.) Social media has become an uncomfortable reminder of how our disconnection from nature—our pollution of rivers, seas, and lakes—will come back to haunt us when we return to it.

While all the negative consequences of disconnecting from nature are undeniable, the positive outcomes of living in clean, safe environments that are closer to our natural origins are worth celebrating and can be very surprising. As a random, joyful example, alongside improved general well-being and physical health, incredible studies by Q Li and several other Japanese researchers found that forest bathing (spending time mindfully with trees) increased the number of anticancer proteins in both women and men and that this increase lasted for seven days after forest bathing.[21] Isn't that incredible?

On a very basic, experiential level, doesn't everyone feel better after spending time in nature? The resounding answer is yes, because we *are* nature. Not having access to it is akin to missing a part of who we are. And our minds and bodies know it. The bottom line is that, in order for us to be holistically healthy and happy, human beings must prioritize space for nature to heal and thrive.

The Role of Nature in Regenerative Placemaking

By now, I hope to have made it clear why nature is the second pillar of our approach to urban design. In the context of regenerative placemaking, nature is not an afterthought; it is a cornerstone. We must have access to it on a daily basis. It's a lifeline for us all, as well as a birthright.

Do you live in a place that's intertwined with and surrounded by greenery, plant life, trees, and other animals? Do you grow your own food? Do you live off the land, turning to the wisdom of nature to revitalize your mind and body? If you do, I congratulate you, because you're in the minority.

In our modern environments, we often find ourselves surrounded by vast expanses of concrete and steel, living and working in high-rise buildings and sprawling urban landscapes. These settings, while convenient, starkly contrast with the natural world from which we evolved and are in desperate need of rewilding. The concrete jungle of our cities, with their often relentless grid of roads, sidewalks, and skyscrapers, can be visually overwhelming and mentally and physically taxing for even the most modern human. We were simply not designed to live in urban sprawl. Our modern environments are characterized by their high-paced, sedentary lifestyles too. We spend long hours indoors, disconnected from the natural elements that were once integral to our survival (and still are). Green spaces, when they do exist, are frequently confined to small parks or manicured gardens, serving more as aesthetic enhancements than as essential components of our daily lives.

That's where regenerative placemaking comes in. Regenerative placemaking knows the value of living spaces that are not just harmonious with nature but intrinsically connected to it.

The Values of Nature in Regenerative Placemaking

Our pillar of nature stands for certain values, and these values will guide us in any regenerative placemaking project.

Value 1: Sustainability Isn't Enough

Regenerative placemaking shines a light on the following truth: Sustainability is not enough. Sustainability aims to maintain our survival and prevent further degradation; therefore, it is the bare minimum practice we should be embracing. Sustainable, or *green*, design usually focuses on doing less harm to the environment, as opposed to avoiding the harm entirely. It's about simply slowing down the damage we do to Earth's natural systems.

Supporters of regenerative approaches believe we need to think bigger. They urge us to move beyond fragmented, short-term solutions and instead embrace a holistic vision that nurtures both people and the planet. True regenerative design doesn't just slow environmental decline; it nurtures ecosystems where nature and humanity flourish together, creating a future that is not just sustainable but thriving, resilient, and abundant for generations to come.[22]

First, we become sustainable in the built environment; then, we become resilient; then, we become regenerative. This is what it means to become more mature as a species.

Looking at the immature survival-of-the-fittest mentality, it's clear that our current way of living isn't yet in alignment with sustainable, resilient, or regenerative models. Ecologist Lawrence Slobodkin likened evolution to a game where the goal is simply to keep playing. As relatively new players on Earth, humans needed to figure out how to be around in the long term, unlike most species, and most species have gone extinct. For those of us designing and building our living spaces, the new challenge is to stop fighting against nature's changes and start working with them. The first rule of regenerative development is to plan for ongoing change.

Elisabet Sahtouris, an evolutionary biologist, believes cooperation is crucial for a species to succeed. She argues that competition often means a species is still young and trying to claim as much as it can. Mature species, on the other hand, work together and form alliances, leading to stable ecosystems. For example, rainforests have developed over millions of years with no single species in control; instead, all species share leadership and work together. This has meant that, up until human intervention, rainforests were completely regenerative. Sahtouris also notes that complex life forms like multicelled organisms are successful because their cells cooperate rather than compete. She thinks that, for any species to thrive, it needs to support and sustain others—something modern humans are just starting to understand.[23] Ancient wisdom, however, has known this for a while now.

For generations, Indigenous communities across the globe have been developing circular ways of living that are incredibly mature. Remember: One of the most well-known Indigenous principles of regeneration is seventh-generation planning. The agreement between all members of the community is that any resources from Mother Nature must be related with in a way that maintains their abundance for at least three future generations. This principle is reflected in native practices such as controlled burning, where dense undergrowth, shrubs, and small trees are purposefully and carefully burned to prevent uncontrollable fires later, a method now adopted by governmental fire management agencies worldwide.

Indigenous land stewardship is also evident in regenerative agriculture practices such as cover cropping and multicropping. In these practices, plants are intentionally grown together to build a healthy nutrient and moisture balance in the soil and to preserve biodiversity. While Indigenous communities have not traditionally tracked nutrient levels, the knowledge of the land built and passed down through generations has given rise to innovative practices of land stewardship that offer significant opportunities for worldwide regeneration movements. This knowledge, known as *traditional ecological knowledge*, has allowed communities to live in harmony with the land for hundreds of years and has become an integral part of modern society.

Traditional ecological knowledge focuses less on why things are and more on how they are. After generations of interactions with the natural world, this form of cultivated information offers a rich database of community research on plants, animals, medicines, weather patterns, geology, and technologies. The deep understanding of the natural world underscores the importance of integrating traditional ecological wisdom into contemporary practices for a sustainable—and then some—future. For anyone who's interested in finding out more, websites like The Indigenous Environmental Network and the traditional ecological knowledge section of The Environmental Protection Agency's site provide valuable additional information and resources. Additionally, the book *Lo—TEK: Design by Radical Indigenism* by Julia Watson delves into Indigenous, nature-based, and inspired designs that are both fascinating and innovative. This work highlights how traditional ecological approaches can influence modern regenerative design.

The bottom line is that regenerative thinking goes a step further than sustainability and resilience by actively improving and revitalizing ecosystems and allowing them to flourish now and in the future.

Value 2: Nature Is Our Guide

Nature is the perfect source of inspiration for creating better, stronger, and more beautiful built environments. As regenerative placemaking grows in popularity, so too does this concept. However, the way natural systems work has led to new and clever design ideas that have already been implemented in many communities.

Take the lotus effect, for example. The self-cleaning property observed in the leaves of the lotus flower inspired the development of advanced surface technologies to be used in urban design. The lotus leaf has a remarkable ability to remain clean and dry due to its unique surface structure, which features microscopic wax-coated bumps that trap air and cause water droplets to bead up and roll off the surface, picking up dirt and debris along the way. This inspired innovations in coatings and materials for a range of uses, including self-cleaning windows, stain-resistant fabrics, and easy-to-clean surfaces for building materials. These technologies are valuable in reducing the need for chemical cleaners.[24]

Another more well-known example is the design of Japan's Shinkansen, or bullet train, which was influenced by the beak of a kingfisher. The train's nose was shaped to reduce air resistance and noise, similar to how a kingfisher dives into water with minimal splash. This design reduces air resistance and minimizes energy use.[25]

Then there's The Eastgate Centre in Zimbabwe, designed by architect Mick Pearce. Pearce's design emulates the natural cooling system found in termite mounds, which maintain a stable internal temperature through a network of ventilation tunnels. By integrating a similar system into the building, Pearce created a structure that uses natural airflow to regulate temperature, significantly reducing the need for air conditioning.[26]

Have you ever seen honeycombs in a building's design? Originally used in the aerospace industry, their hexagonal shapes have been adopted by architects due to their strength and ability to provide excellent structural support while reducing weight, which helps lower construction costs and enhances energy efficiency. The honeycomb structure also allows for better distribution of forces and loads, making buildings more resilient to environmental stresses. This innovative design also optimizes light distribution and looks amazing. It's now a popular choice in modern architecture and can be seen in the following projects:

- The Eden Project (Cornwall, UK): Designed by Sir Nicholas Grimshaw, this complex features geodesic domes with honeycomb-like hexagonal patterns in their structure.
- The Elbphilharmonie (Hamburg, Germany): Designed by Herzog and de Meuron, the Elbphilharmonie's glass structure features a honeycomb-inspired pattern that creates a distinctive reflective surface.

- The Hive (London, UK): Located in Kew Gardens, this structure designed by Wolfgang Buttress is inspired by the intricate patterns of a honeycomb and serves as an immersive art installation that highlights the importance of bees and their role in ecosystems.
- The Al Bahr Towers (Abu Dhabi, UAE): Designed by Aedas, these twin towers feature a dynamic facade inspired by the honeycomb structure. It includes a high-tech shading device that opens and closes in response to the sun.

These inventions and structures are known as examples of biomimicry. Biomimicry, a concept introduced by Janine Benyus in her 1997 book *Biomimicry: Innovation Inspired by Nature*, takes inspiration from nature even further. Benyus defines biomimicry as "learning from and then emulating life's genius," meaning that we use nature's successful strategies, which have developed over billions of years, to solve our own problems.[27] By copying these natural solutions, we can create designs that are both ecofriendly and effective.

Value 3: Repurposing Is King

Repurposing is a powerful strategy in regenerative placemaking that breathes new life into old or discarded buildings and materials, transforming them into valuable new assets with new purposes. This allows us to avoid starting from scratch with entirely new resources. This method, in terms of its environmental benefits, is superior to traditional recycling, which involves breaking materials down into their raw forms—a process that still consumes a lot of energy and resources. Repurposing combines creativity with resourcefulness and keeps aesthetics in mind by focusing on new uses for existing architecture or materials.

At Future of Cities®, my team and the local community joined forces to repurpose what is now The Emerald Station. Here, we took an old, abandoned warehouse and converted it into a vibrant community center. Rather than demolishing the building and constructing a new one, we chose to preserve and repurpose the existing structure. By maintaining the warehouse's original features, such as its weathered brick walls and sturdy steel beams, we not only saved resources but also showcased the building's unique character and historical roots. The result was a beautiful, dynamic space that served the modern needs of the community while honoring its past.

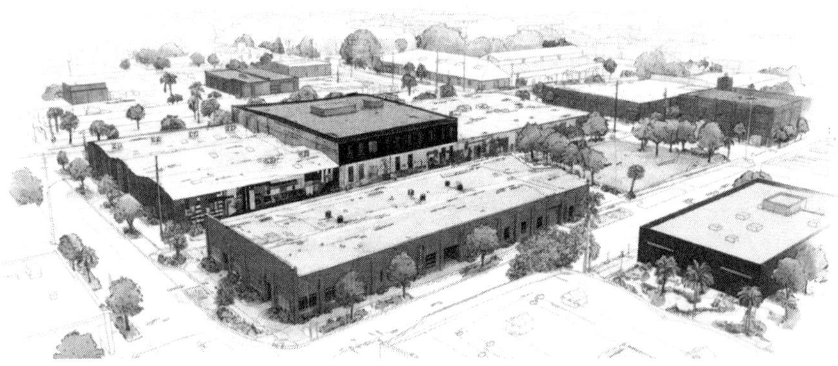

Emerald Station and the Liberty Building at the Phoenix Arts & Innovation District in Jacksonville, FL, are the adaptive reuse of historic industrial structures that anchor the district's creative core. Together they form a commercial hub for local enterprise and community engagement, fostering food and beverage operators, creative industries, and small businesses within a walkable, connected setting. A demonstration project by Future of Cities®

Similarly, Future of Cities®'s Climate and Innovation HUB in Miami stands as another testament to the power of repurposing. For over twenty years, this building has served as a community center, and today, it exemplifies climate resilience, community-centered innovation, and biophilic design. No concrete was used in its renovation from an abandoned warehouse, and it now operates with a ninety-two-kilowatt solar-power system as we work toward an aspirational net-zero energy footprint. Adding to this, we've integrated a stunning reef-inspired installation: a starfish sculpture from the ReefLine project, designed by renowned artist Carlos Betancourt. This is the first prototype of environmentally focused reef art and sculptures, further anchoring the hub's role in showcasing climate solutions and sustainable design.

Another interesting example of repurposing is the High Line in New York City. Originally an old elevated rail line, the High Line was repurposed into a beautiful urban park for the surrounding community to enjoy and for nature to thrive in. Instead of tearing down the rail structure, the project preserved the existing infrastructure and integrated it into a new green space, complete with gardens, walkways, and public art. This repurposing effort not only revitalized an abandoned industrial feature but also created a new community asset.[28]

One of the most innovative examples of repurposing on a grand scale is the Bjarke Ingels–designed ski slope on top of a waste-to-energy plant in Copenhagen. Known as CopenHill, this project transformed an industrial facility into a multifunctional space that serves both the environment and the community. Instead of hiding away this waste management plant, Bjarke Ingels Group integrated a ski slope, hiking trails, and a climbing wall into the design. This ingenious use of space turns what could have been an eyesore into a vibrant recreational area, proving that repurposing isn't limited to small-scale projects.[29]

In the UK, the regeneration of the Battersea Power Station was a huge repurposing success story. This historic coal-fired power station, once a symbol of industrial decline and sitting there derelict for decades, has now been upcycled into a mixed-use development that includes residential, office, and retail spaces for the local and Greater London community. It's one of the most up-and-coming hangouts in London. By preserving the iconic brick facade and core structure, the project maintained the building's historical significance while adapting it for business and pleasure. This approach massively reduced the need for new materials and minimized demolition waste.[30]

Regenerative placemaking goes beyond merely repurposing materials; it's about reimagining and revitalizing spaces to enhance their value and ensure their ongoing contribution to the community's sense of identity. This approach draws on sustainability, resilience, and regeneration, seeing potential in existing structures and materials and transforming what might be considered waste into totally functional assets. This value teaches us that, if in doubt, we should upcycle and repurpose over building anew. Remember this: Don't fix what's not broken. Get creative, reuse, and upcycle.

Value 4: The Greener, the Better

When I refer to *green* here, I mean the color of natural environments within urban built environments. In regenerative placemaking, the principle of the greener, the better highlights the importance of integrating green spaces, parks, living walls, food gardens, permaculture practices, flowers, and trees into urban design. This, of course, beautifies the urban environment and significantly contributes to the community's mental and physical well-being, not to mention the health of the planet.

The "Vertical Forest," the architectural typology that is now recognized as Stefano Boeri Architetti's signature, is the prototype of a new format of architectural biodiversity that focuses not only on human beings but also on the relationship between humans and other living species; Source: Boeri Studio, Milan, 2014; ph. Dimitar Harizanov

An essential aspect of this green integration is rewilding, a process that involves restoring ecosystems to a more natural state by allowing native plants, animals, and ecological processes to flourish. Rewilding can take many forms in urban spaces—such as creating wildlife corridors, restoring waterways, or reintroducing native plant species to city parks and schools. It's a practice that not only adds beauty and biodiversity but also encourages resilience by letting nature regain its natural balance with minimal human intervention. Rewilding wherever possible requires creativity and a commitment to prioritizing nature. And this effort pays off—enhancing air quality, reducing stress, improving community cohesion, improving local biodiversity, and supporting sustainable urban ecosystems.

A prime example of rewilding was brought about by an incredible organization called SUGi. SUGi is a social enterprise founded in 2019 that creates pocket forests in reclaimed urban spaces such as parking lots, schoolyards, and sidewalks. These pocket forests are small, self-sustaining patches of green space that increase biodiversity and strengthen community bonds by encouraging collective ownership. SUGi also proved that these urban green spaces combat the urban heat island effect, which is when concrete and infrastructure trap and store solar radiation, leading to heat-related health issues and diminished air quality.[31] SUGi demonstrated that even small patches of trees along sidewalks can mitigate these effects by providing shade and cooling the surrounding area.

SUGi's impact is international, with nearly 200 pocket forests created across the globe. For example, the pocket forest at MLK Middle School in Berkeley, California, has already improved the students' well-being and learning ability. The trees buffer noise and exhaust from the nearby highway and therefore create a quieter and cleaner learning environment for the kids. They also absorb excess heat, reducing classroom temperatures. The native plants require no watering, eliminating water waste and reducing maintenance costs. (That's another thing to remember as a regenerative placemaker: Native plants are much more ecofriendly than foreign species, so always choose local varieties over imported plants and flowers.)

The success of these pocket forests is largely due to the Miyawaki method, developed by Japanese botanist and ecologist Akira Miyawaki. His forty years of research have proved that forests, no matter how small, using native species, can mature into self-sustaining ecosystems in an incredibly short time period. They're an oasis for biodiversity, healing the environment in various ways and reconnecting local communities with nature and its healing potential. Miyawaki has educated people on planting in over 1,700 areas around the world, including more than 1,400 sites in Japan. This has led to the creation of over 3,000 primary forests and the planting of over forty million native trees worldwide.[32]

The human impacts of SUGi pocket forests are as significant as the ecological ones. Close to 52,000 youth have participated in creating pocket forests in 116 schools. Communities from Chile to Lebanon and from Hong Kong to South Africa and New York have hosted gatherings, including film screenings, educational workshops, and laughter therapy sessions in and around the forests. According to Elise Van Middelem, founder and CEO of SUGi, "The forest acts as an antidote to modern life for humans and wildlife alike."[33]

A similar vision is emerging in London, where a new urban park project is integrating green and blue corridors to create a natural oasis within the city's best neighborhoods. Developed in partnership with the Eden Project, this initiative aims to bring nature closer to urban residents, offering a space for both ecological health and human well-being. I know this because I had the pleasure of meeting Sir Tim Smit himself, the founder of the Eden Project!

Even if there's not enough space for a pocket forest, there is always the option of planting living walls. Living walls are vertical gardens that can be attached to the exterior or interior walls of buildings. These gardens consist of various plant species, often including flowers, herbs, and vegetables, which are grown in soil or hydroponic systems integrated into the structure of the wall. Living walls offer numerous benefits, including improving air quality, reducing urban heat island effects, enhancing the aesthetic appeal of urban environments, and providing insulation to buildings, which can reduce energy consumption for heating and cooling. They also promote biodiversity by providing habitats for birds, insects, and other wildlife. Additionally, living walls can help manage stormwater by absorbing rainwater, reducing runoff, and mitigating the risk of flooding.[34]

Another great example of the positive impact of green infrastructure is seen in Medellín, Colombia. Through extensive efforts in reforestation, urban forestation, and the planting of millions of trees, as well as the implementation of green roofs and living walls, Medellín has successfully reduced its urban temperature by more than two degrees Celsius.[35] This initiative demonstrates how integrating nature into urban planning can lead to substantial environmental and climatic benefits, making Medellín a model city for sustainable development and climate resilience.

Then there's the Million Gardens Project, by Big Green, founded by Kimbal and Christiana Musk. This initiative brings agriculture into urban settings and schools, creating opportunities for students and residents to engage in sustainable food production. By establishing gardens in areas that need them most, the Million Gardens Project not only improves access to fresh produce but also brings environmental awareness to the community.[36] We proudly support this project at Future of Cities®. I was fortunate to connect with Kimbal and Christiana Musk at a fundraiser in Aspen, where we shared ideas on expanding access to green spaces and sustainable agriculture.

In Jacksonville's Phoenix Arts and Innovation District, we're embracing many green initiatives. Plans are underway to install living walls on various abandoned buildings and apartment blocks, to develop as many pocket forests as possible, and to cultivate more community food gardens. We've already made significant strides with our local grow-your-own garden flourishing, but this is just the beginning. We'll also be codesigning the Phoenix leg of the Emerald Train route, incorporating edible landscapes, a food forest, community gardens, play spaces for kids, rooftop gardens, and green walls. These elements will create vibrant, functional green spaces that support both food security and community well-being.

Adopting the principle of the greener, the better is a lifelong commitment, one that involves constant work and development, evolving in tandem with Mother Nature herself. It's a smart, holistic approach that ensures that our designs are not only productive but also reflective of the interconnectedness of nature with all of us.

Value 5: Practice Permaculture

Permaculture is *not* about growing vegetables in cool glass tanks with no soil. (It's fascinating how many people think this!) It's a holistic design framework that extends way beyond sustainable agriculture; it encompasses various aspects of regenerative placemaking, including architecture, social structures, and community development. At its core, permaculture is built on three main ethics: Earth care, people care, and fair share. These ethics guide the creation of sustainable systems that benefit both the environment and society.

Permaculture involves designing landscapes and systems that work in harmony with natural ecosystems. We've embraced this approach at our ChoZen Eco-Retreat and Center for Regenerative Living, for example, where we involve local communities and visitors in educational workshops and experiences that teach practices like homegrown food production, regenerative architecture, and water management. Instead of relying on privatized water that uses chemicals and a lot of energy to clean, we can go for more sustainable options like rainwater harvesting and greywater recycling. These methods help us use natural water resources more efficiently and create greener, more resilient ecosystems.[37]

A bioregional hub is a place-based center that spreads the word about regeneration by integrating local communities, ecosystems, and economies within a specific natural region. Ours is located on forty acres along the St. Sebastian River, adjacent to a 22,000-acre nature preserve that is home to twenty-six endangered species. It serves as a living model for permaculture, and we're very proud of what we're cocreating there. We offer official permaculture ecotours for locals, tourists, and our retreat-goers from around the world. Proceeds from these tours support our nonprofit initiatives, which are dedicated to environmental education and the conservation of the Florida Wildlife Corridor.

If you're interested in visiting us for the day or staying even longer at the retreat center, we'd love to have you. Visit chozencrl.org to find out what's growing next.

The Principles of Permaculture

Permaculture is guided by principles that provide practical strategies for applying its core ethics in various contexts:

- Observe and interact: Spend time observing natural systems and understanding their processes before making changes. Learning from nature ensures designs are responsive and appropriate.
- Catch and store energy: Capture renewable resources, such as sunlight, wind, and rainwater, to store and use for future needs.
- Obtain a yield: Ensure that the systems designed provide immediate or long-term yields, such as food, energy, or other resources.
- Apply self-regulation and accept feedback: Create systems that are self-regulating and adaptable by monitoring their outputs and making necessary adjustments.
- Use and value renewable resources and services: Rely on renewable resources that can regenerate and provide services without depleting the environment, like using compost instead of chemical fertilizers.
- Produce no waste: Design systems that reuse all outputs, turning potential waste into resources, such as composting food scraps.

- Design from patterns to details: Start with a broad perspective by recognizing natural patterns and then work out the finer details for effective design.
- Integrate rather than segregate: Create systems where different elements support each other, leading to greater resilience and productivity.
- Use small and slow solutions: Favor small-scale, gradual interventions that can be adjusted over time, making them more sustainable and manageable.
- Use and value diversity: Diversity within a system increases resilience against pests, diseases, and other challenges, allowing different elements to fulfill multiple functions.
- Use edges and value the marginal: The edges or intersections of different environments often have the most productivity and diversity. Valuing these areas can provide unexpected benefits.
- Creatively use and respond to change: Adapt to change creatively and proactively, turning challenges into opportunities for innovative solutions.

One of the most well-known examples of permaculture in action in the United States is the Beacon Food Forest in Seattle, Washington. This seven-acre community-driven project transformed an underused urban space into a completely edible landscape. Using permaculture principles, the food forest includes a variety of fruit and nut trees, berry bushes, and vegetable gardens. It serves as a community gathering space, an educational resource, and a source of fresh produce for local residents.[38]

Overall, permaculture is a comprehensive approach to designing sustainable and regenerative communities. By integrating its principles into urban planning and community development, we can create places that support ecological health, social well-being, and economic stability, cocreating a more regenerative and equitable future for everyone.

Value 6: Pollution Is Not Inevitable

In regenerative placemaking, the belief that damaging pollution is an unavoidable byproduct of urban development is firmly challenged. Instead, we strive for solutions that are net zero, carbon neutral, or, at best, climate positive, ensuring that our urban environments contribute positively to the planet rather than detracting from it. For those of you unclear about the differences among carbon neutral, net zero, and climate positive, here's a quick breakdown:

Carbon neutrality means balancing the amount of carbon dioxide emitted with an equivalent amount of carbon offsets or reductions elsewhere. Essentially, it's about ensuring that the total carbon emissions from a particular activity or organization are offset by actions that reduce or remove an equal amount of carbon from the atmosphere. For example, a city might achieve carbon neutrality by investing in renewable energy projects or planting trees to absorb carbon dioxide.

Net zero, on the other hand, goes a step further. It involves reducing greenhouse gas emissions to the point where any remaining emissions are counterbalanced by removing an equivalent amount of greenhouse gases from the atmosphere. This concept encompasses all greenhouse gases, not just carbon dioxide, and aims for a balance where the total emissions produced are equal to the total emissions removed or offset. Achieving net zero often involves a combination of reducing emissions as much as possible and using advanced technologies or natural processes to remove any residual emissions.

While both terms are focused on reducing pollution, carbon neutrality is about offsetting emissions to achieve a balance, whereas net zero aims for a complete equilibrium between emissions produced and emissions removed. This commitment to net-zero design not only addresses the immediate impacts of urbanization but also sets a precedent for future developments. Even if we fall short initially, aiming for net zero and landing at carbon neutral still represents meaningful progress—a bit like shooting for the moon and landing among the stars.

A major part of this journey is the shift toward regenerative agriculture, which has gained momentum thanks in part to the documentary *Kiss the Ground*. This film has catapulted the regenerative agricultural movement to the forefront of our thinking around the food industry, shedding light on industrial agriculture as a major contributor to climate change and the current health crisis.[39] My team and I proudly support *Kiss* the *Ground* and its founder, Ryland Engelhart, who has played a key role in influencing policy and spreading awareness about the benefits of regenerative practices.

Another inspiring voice in this movement is Mark Hyman, a prominent functional medicine physician actively lobbying in Washington, DC, to reform the US food system, which he argues is harming public health. Both Engelhart and Hyman have used their platforms to drive meaningful change, and their work underscores the importance of a holistic approach to environmental and human health.

This holistic approach also guides our urban infrastructure projects. Take some of the green spaces planned within the Phoenix Arts and Innovation District that are now officially part of the Emerald Trail. Once the bike lane system is complete, it will feature over thirty miles of trails, greenways, and parks, linking downtown with fourteen historic neighborhoods and key landmarks such as Hogan's Creek, McCoy Creek, and the St. Johns River. The trail will connect sixteen schools, two colleges, three hospitals, twenty-one parks, and the Regional Transportation Center, along with numerous restaurants, retail spaces, and businesses. Groundwork Jacksonville is leading the effort, coordinating with us at Future of Cities® and other stakeholders to leverage resources and realize the trail's full potential to dramatically reduce pollution.

One of the most exciting aspects of the Emerald Trail is its goal to create equity by linking underserved downtown communities, as well as providing them with access to affordable, healthy food. This connectivity is also expected to reduce health risks, boost happiness, and bring about equity in neighborhoods that have long been overlooked. Mayor Donna Deegan has made this area a priority, recognizing the Emerald Trail as a way to address disparities and uplift these communities. This project won't just cut down pollution from car traffic but will also stimulate social and economic growth, enhance public safety, and contribute to a more vibrant, inclusive Jacksonville.[40]

But projects like the Emerald Trail with its bike lanes are just one piece of the puzzle. The integration of living walls, pocket forests, and urban gardens provides additional avenues for offsetting carbon. Green roofs and walls, for instance, absorb carbon dioxide. They also offer a natural cooling solution that reduces the need for air conditioning, thereby lowering energy consumption and, therefore, carbon emissions.

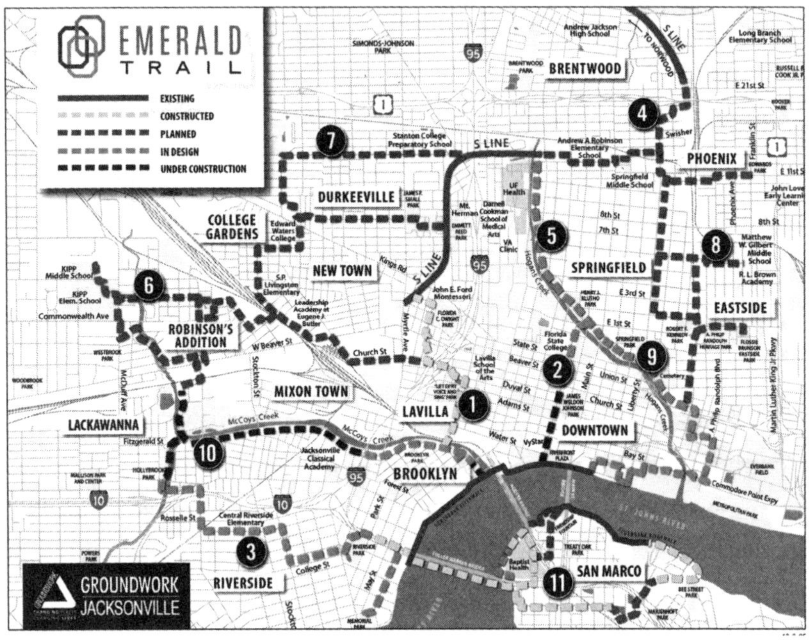

Map of 30-mile Emerald Trail showing half of the trail in design, under construction, or complete by Groundwork Jacksonville

When we're actively removing pollution, that's when we reach the third level of environmental cleanliness, the gold standard of *climate positive*. Climate-positive actions don't just "do no harm"; they actively help reverse the damage we've done as a collective. It goes beyond net zero and carbon neutral to actively remove additional carbon dioxide from the atmosphere, thereby having a net beneficial impact. Being climate positive, also sometimes referred to as *carbon negative*, means that a person, organization, or project not only balances out its carbon footprint but also contributes to the overall reduction of greenhouse gases in the atmosphere. Community-run permaculture farms like the one I mentioned at the ChoZen Eco-Retreat and Center for Regenerative Living, as well as the wetlands surrounding them, are examples of climate-positive spaces. They act as carbon sinks, capturing carbon dioxide and cleansing the atmosphere.

Another innovative climate-positive solution is the incorporation of carbon capture technologies in building designs. These technologies are designed to capture and store carbon emissions before they enter the atmosphere, making it possible to retrofit existing buildings with systems that actively work to reduce their carbon impact.

At best, we're cleaning up the mess we've made in terms of pollution, and at a bare minimum, we're not adding to the problem. That's where the choice of materials we use in urban design comes in. Repurposing is king, with the use of preexisting materials reducing a lot of pollution. In regenerative placemaking, the use of energy-efficient windows, insulation, and new ecofriendly building materials (when new materials must be used) all minimize pollution.

Arguably, the most powerful tool for cutting carbon emissions and pollution on a global scale is renewable energy—not to mention its potential to create a wealth of new jobs. As Chris Castro, adviser to Future of Cities® and chief of staff for the former secretary of energy, Jennifer Granholm, put it, "The transition to a clean and renewable energy future is the greatest economic and job creation opportunity of the twenty-first century."[41] Integrating renewable energy into urban design could take many forms, such as these:

- **Solar power**
    - **Rooftop solar panels:** Install photovoltaic panels on rooftops of residential, commercial, and public buildings to harness solar energy.
    - **Solar canopies:** Use solar canopies over parking lots, walkways, and public spaces to generate electricity while providing shade and protecting vehicles.
    - **Building-integrated photovoltaics:** Integrate solar panels into building materials, such as facades and windows, to blend energy generation with architectural design.
- **Wind energy**
    - **Urban wind turbines:** Install small-scale wind turbines on rooftops or in open urban spaces where wind conditions are favorable.
    - **Vertical axis wind turbines:** Use these turbines in urban environments where space is limited, because they are often more efficient in turbulent wind conditions typical of city settings.
- **Geothermal energy**
    - **Geothermal heating and cooling:** Implement geothermal heat pumps for heating and cooling buildings. These systems use Earth's constant temperature to regulate indoor climates efficiently.
    - **District heating systems:** Develop geothermal district heating systems for neighborhoods or districts, providing a centralized source of renewable energy for multiple buildings.

- **Biomass and biogas**
  - **Biomass boilers:** Use biomass boilers to generate heat and power from organic waste materials. This can be integrated into community buildings or residential complexes.
  - **Anaerobic digesters:** Implement anaerobic digesters to convert organic waste into biogas (these are like small spaceships that make waste into biofuel), which can be used for electricity generation or heating.
- **Hydropower or hydroelectric power**
  - **Microhydro systems:** Small-scale hydroelectric systems can be installed in urban waterways, such as rivers or canals, generating renewable energy while minimizing environmental impact.
  - **Run-of-river hydropower:** This type of hydroelectric power, where turbines use the natural flow of water without large dams, provides a sustainable option for areas with sufficient water flow.

The list goes on. Let's not limit our options or imagination here. When we harness the full spectrum of renewable resources, ambitious sustainability goals become very achievable. While not renewable, nuclear and fusion power also deserve a mention as low-carbon energy sources that some countries rely on to reduce their emissions. Although we didn't include them in our list, they remain significant options for generating clean energy and have the potential to support the transition to a low-carbon future. When we start embracing carbon-neutral and net-zero solutions, we'll be paving the way for cities that not only function efficiently but also actively contribute to a healthier planet. Pollution should not be treated as a byproduct of the existence of humans—period.

For those who are in doubt as to whether any of this matters and whether we can really make a difference anyway, just cast your mind back to 2019. The pandemic underscored the impact of human activity on the natural environment and its ability to regenerate itself. Despite the severe effects on populations and economies, it revealed how changes in urban living could benefit our ecology.

During the pandemic lockdowns, global carbon dioxide emissions dropped by approximately 6.4% in 2020 from 2019 levels, the largest annual decrease ever recorded. Improvements in air quality were also significant: Cities like New Delhi saw a reduction in PM2.5 concentrations by as much as 60%. Additionally, satellite imagery showed cleaner waterways and a decrease in nitrogen dioxide levels in major industrial areas.[42] These glimpses of nature's resilience remind us of what's possible when human activity decreases and ecosystems are given a chance to rebound.

How the Pillar of Nature Meets the Pillar of Community

As you make your way through this book, you'll increasingly come to understand that no living system or pillar of regenerative placemaking stands alone. This is especially true of the community and nature pillars.

Regenerative placemaking hinges on the understanding that creating vibrant, resilient, and ecofriendly urban spaces requires a deep integration of community involvement. To facilitate this becoming a reality, and not just a beautiful idea, a comprehensive plan is essential—one that guides design, development, and constant learning for everyone involved.

For example, a simple community garden, with flowers, fruits, and vegetables, doesn't just benefit the planet and massively reduce pollution. It benefits the mental and physical health of the community by serving as a meeting spot for neighbors to work together and build relationships. It benefits the community's physical health by improving their fitness (gardening can be hard work!) and encouraging healthy eating, and their mental health skyrockets as a result. For example, one study by the University of Maryland showed that the simple act of tending to a garden leads to improvements in mental health, including reduced stress and enhanced mood.[43]

Investment in green spaces, walkability, cyclability, biodiversity, permaculture, renewable energy, and natural capital significantly enhances a community's health and well-being. And nature itself, through its evolutionary wisdom, presents dynamic systems of unmatched efficiency and should be leaned upon for inspiration.

The success of this pillar depends on effective education for the community of any given area to be regenerated. Transdisciplinary education plays a vital role in facilitating the exchange of knowledge necessary for regenerating both built and natural environments. Regenerative placemaking interventions, such as pop-up parks, workshops, and festivals, provide opportunities to test and showcase educational resources that can have far-reaching impacts over time. These events help support a very important shift in mindset toward prioritizing nature and underscore the need to integrate ecofriendly systems into our communities, for the sake of those communities. Only education can translate how connecting with nature and adopting sustainable practices can enrich our lives and our surroundings. Through these educational efforts, we lay the groundwork for a more informed and fully engaged community that values and actively supports environmental regeneration, to the point where communities are willing to invest time, energy, and resources into initiatives that support the pillar of nature.

Take the town of Minster, Ohio, whose community embraced a community solar program to make renewable energy more accessible to all its residents. Partnering with American Municipal Power, Minster developed a 4.3-megawatt solar array with a 7-megawatt battery storage system. This initiative allowed the residents to rally together to create a shared solar array, reducing their energy bills, enhancing local grid resilience, and reducing their pollution. The town's commitment to sustainability by lowering its carbon footprint and supporting local renewable energy sources was brought about through educational initiatives and cross-collaboration. Minster's local government and village administrator Don Harrod played a pivotal role in championing the project, advocating for its potential to provide economic, environmental, and energy security benefits. The initiative was presented to the community in organized meetings that outlined the benefits for both people and planet. The community showed a willingness to invest time, energy, and resources into the project, reflecting the town's commitment to the regeneration of local ecological systems.[44]

It is community education and engagement that drives substantial environmental progress.

Hope Is on the Horizon

Stories like this support the notion that, as a species, we are maturing. Prioritizing nature and protecting our land, waters, air quality, and biodiversity for the good of the collective is a symptom of that growth. This fills me with hope, and I hope it does that for you too. Our species has undergone huge change over our time on Earth, from the Bronze Age, Iron Age, and ice ages to the digital era, from world wars to natural disasters. We've constantly adapted to immense changes. Now, we face yet another challenge on a global scale: climate change.

But just as we've navigated past crises, I believe we can rise to this one as well. As Charles Darwin suggested in his theory of natural selection, it's not necessarily the strongest or most intelligent species that survive but those most responsive to change, and, I'd argue, responsibility. By embracing regenerative placemaking, we have the power to slow the pace of climate change and build a better future together. There is hope if we embrace our power and influence for good.

This reminds me of an old Sufi story. In a village, there was a man known far and wide for his wisdom. Two young men, eager to test his reputation, devised a plan. The conversation went something like this: "Let's catch a small bird," one said. "We'll ask him whether it's alive or dead. If he says it's alive, I'll crush it in my hands. If he says it's dead, I'll release it, making him wrong either way!"

They approached the wise man, hiding the bird in their hands, and asked, "Wise one, can you tell us if the bird we hold is dead or alive?" The sage looked at them with calm eyes, smiled gently, and replied, "It's in your hands."

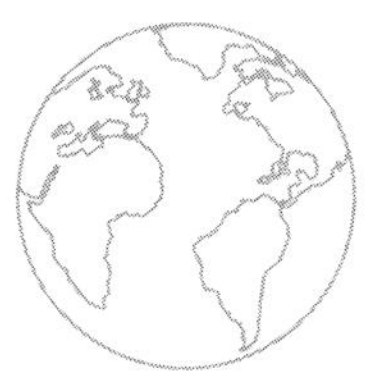

CHAPTER 5

Celebrating the Soul of the City

Culture is the widening of the
mind and of the spirit.[1]

—JAWAHARLAL NEHRU

Have you ever tried to define your own culture or explain the concept of culture itself? Despite being an integral part of our daily lives—something we are born into and experience every day—culture is so intrinsic to us, so close to us, and so much a part of who we are that it can render itself invisible. Understanding it entirely can be as elusive as trying to see your own eyes, but as regenerative placemakers, we should understand what it truly means so we can best serve our communities.

One reason why defining culture can be so challenging is that it encompasses a broad spectrum of experiences and practices, intersecting with numerous subcategories. Before we get into culture's role in regenerative placemaking, let's explore what it's about.

The Limbs of Culture

In general, culture encompasses the collective practices, beliefs, values, customs, and artifacts that characterize a group of people. Moreover, it shapes and reflects the way individuals within a community interact with each other and their environment. Any definition of culture must include the following:

Beliefs and Values

Core beliefs are fundamental ideas that guide a community's behavior and decision-making within a culture. For example, beliefs about what is right or wrong or what constitutes a good life shape how people interact with each other and with their environment.

Religion often provides a framework for understanding existence, ethics, and purpose. It influences rituals, moral codes, and community practices. Different cultures may have varying beliefs about the divine, spirituality, and the afterlife.

Moral values are the standards by which people judge what is good or bad behavior. They can encompass ideas about honesty, justice, loyalty, freedom, love, compassion, and respect. Moral values often inform laws and social norms. For example, in the United States, the moral value of individual freedom is paramount. This is reflected in laws that protect free speech.[2]

Customs and Traditions

Rituals and ceremonies are formalized practices that mark significant events or transitions, such as weddings, funerals, or rites of passage. They often involve specific customs and symbolic actions that hold cultural significance.

Festivals and celebrations are periodic events that bring people together to celebrate cultural heritage or religious observances. Examples include Diwali in India, Carnival in Brazil, and Thanksgiving in the United States.

Everyday practices such as mealtimes, greetings, and ways of expressing hospitality reflect cultural norms and values.

Language and Communication

Different languages can shape thought patterns and cultural perspectives. For example, the word *saudade* in Portuguese doesn't have a direct translation in English, but it's along the lines of "melancholic longing," "yearning," "nostalgia," "loneliness," and "incompleteness." It's usually used in a romantic or sentimental context about a lost loved one. This strange and beautiful term went on to influence an authentically Portuguese music genre known as *fado*, which brings together haunting singing with stirring, mournful tunes.[3]

Nonverbal communication includes body language, facial expressions, and gestures. Nonverbal cues often convey meanings that are context dependent and culturally specific. For example, Italian hand gestures have over 250 meanings![4]

Cultural symbols such as flags, logos, and religious icons represent and communicate cultural values and identity. For example, the Japanese flag, called the *Hinomaru*, features a red circle representing the sun and symbolizes Japan's connection to nature. Known as the Land of the Rising Sun, Japan reveres the sun goddess Amaterasu, reflecting the nation's pride and respect for tradition.[5]

Arts and Literature

Painting, sculpture, and architecture reflect cultural aesthetics, historical events, and social values. Artistic styles and techniques can signify cultural and historical periods. For example, the Renaissance art movement, starting in fourteenth-century Italy, is celebrated for its focus on humanism and naturalism, as seen in Michelangelo's *David*. This piece highlights the importance of individualism in society at the time.[6] Compare that to the street art you can see in the Phoenix Arts and Innovation District in Jacksonville, and you'll see a notable difference in the themes explored by our artists, such as freedom, creativity, resilience, and togetherness.

Music and dance convey emotions, tell stories, and reinforce cultural identity. Traditional music and dance often play a role in rituals and celebrations. These vary massively in cultures. Just compare Latino reggaeton with British ballroom dancing!

Written works such as novels, poetry, and folklore preserve and transmit cultural narratives, historical events, and societal values. Literature can profoundly influence a nation and, indeed, the entire world. Just think about the impact of texts like the Bible, the Quran, and the Torah, which have shaped cultural identities across the globe.

Social Norms and Practices

Social norms define expected behaviors in various situations. This includes how people should interact, dress, and behave in public and private settings. What you'd wear in Miami Beach in the height of summer would probably be quite frowned upon if you wore the same clothes, or lack thereof, in the city of Mecca.

Cultures have different ways of organizing society, including class systems and family structures. These structures influence power dynamics and social interactions.[7]

Food and Cuisine

The type of food commonly consumed, preparation methods, and culinary traditions reflect cultural identity. Cuisine can include regional specialties, staple foods, and unique cooking techniques. For example, in Japan, you have sushi with its emphasis on fresh, raw ingredients and precise preparation, and in Canada, you have poutine (double deep-fried french fries topped with cheese curds and smothered in gravy) with its emphasis on . . . deliciousness and diabetes?

The ways in which food is shared and consumed, including meal times, dining etiquette, and communal eating practices, can signify cultural values and social bonding. If you want to get some clues about a place, try reading a local cookbook.

Certain foods hold symbolic meaning or are associated with specific occasions, such as festive foods or ritual offerings. For example, during the Jewish holiday of Rosh Hashanah, apples dipped in honey are traditionally eaten to symbolize a sweet new year.[8] Similarly, during the Chinese New Year, dumplings are often served because their shape resembles ancient Chinese gold ingots, symbolizing wealth and prosperity.[9]

Technology and Innovation

Cultures adapt to and integrate new technologies in ways that reflect their values and priorities.

Cultural attitudes toward innovation and progress can significantly shape how new ideas are either embraced or resisted. Different cultures prioritize various types of technological advancements based on their values and needs. For example, in Japan, a culture that highly values technological innovation, there is strong support for advancements in robotics and digital technology.[10] In contrast, Scandinavian countries, with a strong emphasis on environmental sustainability, prioritize ecological technologies, such as renewable energy solutions and green building practices.[11]

Economic and Social Structures

The way a society organizes its economy—capitalist, socialist, mixed—reflects its cultural values and priorities regarding wealth distribution, work, and consumption. The United States is well known for its capitalist values, China for its communism, and Cuba for its socialism.

Institutions or systems such as family, education, and government reflect and reinforce cultural values and norms. They play a huge role in shaping social behavior and community life.

How people form and maintain relationships within their communities can vary widely. Social networks and community support systems are influenced by cultural norms. For example, in South Korea, there is a strong emphasis on close-knit family units, respect for elders, and hierarchical relationships.[12]

Identity and Heritage

Cultural perception encompasses how individuals and groups perceive themselves and are perceived by others based on cultural background, ethnicity, and heritage. For instance, in multicultural societies like Canada, individuals often develop a hybrid cultural identity that integrates elements from various backgrounds. A Canadian of Asian descent might celebrate both traditional Asian festivals and Canadian national holidays, blending heritages to form their unique cultural identity.[13]

Efforts to preserve and celebrate cultural heritage include maintaining traditional practices, languages, and historical sites. Heritage initiatives help keep cultural knowledge and identity alive.

Culture is dynamic and evolves over time. The interplay between preserving traditions and adapting to new influences reflects a culture's response to internal and external changes. In Japan, traditional practices such as tea ceremonies and wearing a kimono continue to be highly valued and preserved. However, modern Japanese society has also embraced contemporary influences, such as Western fashion and technology. For instance, while the traditional kimono remains an important cultural symbol, many Japanese people wear Western-style clothing daily. This blend of old and new illustrates how Japanese culture maintains continuity through its traditions while adapting to global changes and modern lifestyles.[14]

As you can see, culture is a rich and complex concept that profoundly shapes human experiences and interactions. It is not just a static set of traditions or historical artifacts but a dynamic, evolving system that grows and transforms over time. Culture is deeply rooted in the story of place, shaped by the unique historical paths societies have traveled and the interconnected lives of human and nonhuman descendants of that place. These descendants—people, plants, animals, and ecosystems—are essential to understanding the essence of both place and culture. At the same time, culture is a living entity, constantly adapting and responding to new influences, innovations, and global interactions.

While culture plays a crucial role in shaping how societies function, thrive, and maintain their identities, not all culture is healthy. A healthy culture is based on inclusivity, mutual respect, and collective well-being. However, some cultural practices or norms cause harm, inequality, or oppression and shouldn't be supported by the masses; they require huge critical reflection and reform. For example, certain gender norms in various cultures may limit opportunities for women or reinforce abuse and oppression. Similarly, many cultures marginalize racial and ethnic minority groups, sometimes subtly and sometimes loudly, whether through exclusion, outright discrimination, or violence. Addressing these negative aspects of unhealthy branches of culture head-on is essential if we want to nurture cultures that promote fairness, dignity, and the flourishing of all members.

Culture is both a reflection of past experiences and a guide for how communities navigate present and future challenges. It's safe to say that culture plays one of the biggest, most influential roles in how societies and communities function, thrive, and maintain their identities amid a changing world.

Why Culture Is the Third Pillar of Regenerative Placemaking

Protecting, preserving, and enhancing culture is crucial in regenerative placemaking for several interconnected reasons. For one, culture forms the backbone of a neighborhood's, city's, or country's identity and heritage, which makes this pillar interdependent with the first one: community.

From a Regenesis perspective, the story of place and the vocation of place are central to its culture and history. The story of place captures the unique history, nature, and essence of a place and the larger system it is nested in. The story acts as a narrative, using metaphors rather than just dates and events, to highlight what makes a place different. It helps us see the whole system as the context for solving issues in a way that fits the living environment, rather than imposing solutions from elsewhere that may not work well with this specific culture and ecology.

The vocation of place refers to its deeper role within both the nearby and greater wholes in which it is nested. These wholes could be local ecosystems, regional cultures, or broader social and environmental networks. By understanding a place within these larger contexts, regenerative placemaking ensures the place stays culturally aligned with its purpose and connections to the broader systems to which it belongs.

And as a regenerative placemaker, no matter what your role is, when you help maintain and celebrate a community's history and culture, you show them respect. There's no greater act of reverence for a community than fighting to represent its unique identity and historical narrative in your regenerative placemaking initiatives.

In this process, it is important to recognize that culture, diversity, and history are deeply interrelated. Just as the most resilient ecosystems in nature are those that are diverse—composed of synergistic and complementary species—so too are human communities. As we have seen, in nature, a *guild* refers to a group of different species that interact and support one another within an ecosystem, each playing a unique role while collectively enhancing the system's overall health. Similarly, human guilds function when diverse groups come together, each contributing their distinct skills to strengthen the social ecosystem.

In regenerative placemaking, guilds are more than just trade associations; they are dynamic ecosystems where diverse groups collaborate to support one another. At PHX-JAX, the Climate and Innovation HUB, and the ChoZen Eco-Retreat and Center for Regenerative Living, different cultural and professional communities work together like members of a guild, cocreating opportunities for connection, learning, and sustainable growth.

To ensure long-term impact and financial sustainability, our business model integrates mission-driven programs with diverse revenue streams—from donations and sponsorships to merchandise sales, local agriculture, educational offerings, and artist residencies. This holistic approach benefits individuals as well as entire communities.

Just as biodiversity strengthens ecosystems, cultural and economic diversity fuels regenerative development. By embracing different experiences, talents, and perspectives, these interconnected guilds form a strong, adaptive microculture, proving that true regeneration happens when communities uplift and empower one another.

Holding a community's cultures in high regard is a surefire way to get the community members on board, engaged, and supportive of your endeavors too. After all, success moves at the speed of trust, and you build a lot of trust by respecting a community's culture, opinions, and ideas when it comes to the regenerative placemaking project.

A recent example in Florida shows what can happen when a community's culture isn't involved in the planning process. The Florida Department of Environmental Protection announced plans to develop nine state public parks with golf courses, pickleball courts, hotels, and glamping sites. This proposal, part of the 2024–2025 Great Outdoors Initiative, was announced with only a week's notice for public input.[15] It became obvious that many of the locals weren't generally the type of people who golfed and glamped, and they didn't want their precious green spaces spoiled by unwanted amenities. The backlash was immediate and strong, leading the developers to drop the proposals within a week.[16]

This highlights how important it is to involve the community from the start in any placemaking initiative; ignoring public input can quickly erode trust and derail well-intentioned projects. Real regenerative placemaking works best when it's a collaboration with the people it affects, making sure their voices are heard and their needs are respected.

When cultural values and traditions are reflected in the built environment, residents are also more likely to feel a sense of ownership and responsibility toward these built or natural areas. If a community in Jacksonville rallied together to codesign a cultural hub in a building with historical significance and one that was a symbol of the community's resilience, they're much more likely to help in the upkeep of that building. If a community in Paris helped plant the tender saplings of a pocket forest with native French trees and flowers, they're more likely to be invested in the preservation of that area, making sure it's clean and respected. This results in much more loyalty and effective stewardship of the areas you are regeneratively developing. That's a win–win for everyone and ensures the longevity of a project's success.

The opposite is also true. Neglecting or dismissing a community's cultural heritage can have profound consequences, including the erosion of trust, disengagement, and conflict. A notable example is the 2004 unrest in Kosovo, where ethnic tensions led to the destruction of numerous Serbian Orthodox churches and monasteries by ethnic Albanians. This deliberate targeting of cultural and religious sites was perceived as an attempt to erase Serbian cultural identity, resulting in widespread violence and displacement of communities. Such incidents underscore the critical importance of respecting and preserving cultural heritage to maintain community trust and cohesion.[17]

But back to some positive examples. Consider the importance of cultural preservation in the context of Miami Beach's Art Deco Historic District. Imagine how different Miami Beach would look today if the city had allowed developers to demolish the architectural treasures of the art deco era. Would Miami Beach have become a sprawling, high-rise landscape like Las Vegas? Instead, the city's low-rise, human-scale design has made it one of the most densely populated and walkable cities in the US, alongside New York City and San Francisco. Despite its density, Miami Beach has maintained a small-town vibe, free of towering skyscrapers, which adds to its unique charm. This preservation was made possible thanks to the tireless efforts of the conservationist movement led by Barbara Baer Capitman, supported by figures like Craig Robins and Tony Goldman, who recognized the cultural and historical significance of the area and fought to protect it. Without this movement, Miami Beach could have lost its iconic character.

Remember this: Culture is the main source of a community's sense of belonging and pride, which creates the base for any successful regenerative placemaking project. The protection, preservation, and celebration of cultural heritage also play a significant role in driving economic activity and tourism. Well-preserved cultural sites and traditions can attract visitors from across the region, state, country, and even the world.

Cultural preservation also provides educational opportunities for community members, visitors, and tourists. By showcasing cultural traditions, histories, and practices through public art, architecture, and events, regenerative placemaking educates people about the rich heritage of a community. The regenerative placemaking approach honors the past, lives in the present, and cocreates the future *with*, *by*, and *for* the community. By making culture central to placemaking, we ensure that spaces feel meaningful and personal to those who live there—truly like *home*. As a result, culture stands as a crucial pillar in our regenerative placemaking framework, alongside community and nature, shaping the design and innovation behind every project.

Integrating Culture: What Does It Look Like?

In regenerative placemaking, the pillar of culture goes far beyond stereotypical aesthetics, beauty, or surface-level decoration. It's not just about hanging lanterns in Vietnamese projects or hosting tea parties in British ones. At this stage of planning, we must resist clichés, generalizations, and assumptions about what is culturally appropriate. More importantly, it's about embedding cultural narratives, histories, practices, and values into the development process.

It's also about recognizing the darker side of a place's history—cultural whitewashing, ethnic cleansing, and the forced removal of communities and their traditions by those in power. For example, traditional celebrations like Columbus Day overlooked the experiences of Indigenous peoples and the suffering caused by colonialism—although today, it is increasingly recognized as Indigenous Peoples' Day in an effort to honor those who came long before European settlers.

Many developers, often unknowingly, act as modern-day colonizers by imposing their own ideas of what is culturally appropriate or beneficial for a community, without truly codesigning projects alongside the people who live there. True regenerative placemaking moves away from this approach, instead focusing on collaboration with the community, incorporating nature, culture, and local voices into every project to ensure that the identity of a place is honored and preserved.

Regenerative placemaking should not only protect but also enhance a place's culture. A way to do this is to create niches where people feel they belong—places where they can be in their culture and share their culture with others. This might include incorporating local art, celebrating cultural festivals, and reflecting the values and histories of the community in the architecture. In Scandinavia, for example, especially in Denmark and Sweden, regenerative design is based on a strong cultural commitment to nature, simplicity, and functionality. Danish *hygge* and Swedish *fika*, translating roughly to English as "a quality of coziness and comfortable conviviality," are cultural practices that influence the design of public spaces, encouraging designers to incorporate a sense of comfort, sociability, and warmth into their projects. This is seen in the integration of cozy communal areas, natural materials, and functional yet aesthetically pleasing design elements.[18]

Culture is highly dynamic, evolving over time while simultaneously retaining core traditions and values. In the context of regenerative placemaking, acknowledging this duality and cocreating environments that honor the past and embrace the future are vital. For instance, Kyoto, Japan, reflects this balance with its blend of traditional wooden *machiya* houses and modern, ecofriendly developments alongside them. This approach preserves the city's historical charm while supporting sustainable living practices and the modernity that Japanese society so famously embraces.[19]

Regenerative Tastemaking

Food is more than sustenance and music is more than sound—they are full of rich stories and histories, creating gathering places as a celebration of culture. Across the world, cuisine and music are a shared language that invites us into the heart of a community, allowing us to taste its traditions, values, and creativity. Whether around a family table or a buzzing neighborhood spot, food and music connect us beyond borders—offering a tangible, flavorful and rhythmic expression of identity and belonging.

One standout example of this ethos is Grassfed Culture Hospitality Group, a US-based culinary collective with a mission "to build a future where urban life is not only sustained but reimagined—one that champions regeneration, enriches culture, and enhances the human experience."[20] Their approach isn't just about serving meals or programming musical acts —it's about crafting meaningful cultural experiences.

Through concepts like ZeyZey in Little Haiti and the Michelin-Star and Green Star Krüs and Los Félix in Coconut Grove, they're reimagining what it means to dine and dance in Miami. These aren't just restaurants or a music venue—they're living expressions of place and purpose.

"In a city known for the fleeting and the flashy, ZeyZey pulses at the heart of counterculture—a sanctuary for global sound rooted in soul, rhythm, and rebellion."

—Tigre Sounds[21]

More than meals or performances, Grassfed Culture offers regenerative journeys into place—where multicultural hospitality and culinary excellence come together to nourish both people and the land.

This idea of regenerative tastemaking is also beautifully illustrated by a company called Botmas in the Netherlands (Botmas.nl), a farm that grows over 400 cultivars, inviting children to taste them and invent recipes they can share with their families. Botmas even organizes cooking sessions in local care homes, where older residents teach traditional recipes using local ingredients, bridging generations through food.[22] This approach to food as a bridge between ages and cultures is a fantastic model for building community resilience.

For regenerative placemakers, food isn't just important—it's *the secret sauce*. So don't forget to include tastemakers as part of your strategy to bring people together and celebrate your place's culture.

Arts and Events

In policy planning and urban studies, I've often seen the role of arts and culture receiving secondary attention compared with more pressing urban living concerns like transportation, infrastructure, housing, and crime reduction. Arts and events are frequently siloed into the entertainment industry, which completely overlooks their integration into and interconnection with many aspects of urban life for any community, including its economy and the quality of life, mental health, and all-around social well-being of its people. Studies have shown that participation in cultural activities, such as attending cultural events or engaging in artistic activities, is positively associated with higher levels of happiness and life satisfaction.[23] At both micro and macro levels, the arts play a central role in a city's development success.

Can you imagine a city or neighborhood that everyone wants to live in and that is devoid of good food, music, art, history, and events and activities for people of all ages? Culture expresses itself through these avenues. Culture is the reason we call a country, city, or neighborhood home. Similarly, culture is why we travel to other places too (and contribute hugely to the local economy). Without culture, a city is in danger of becoming spiritless. Yet funding for the arts continues to diminish due to a chronic lack of prioritization in the US and the world over.

In the United States, the budget for the National Endowment for the Arts (NEA) has faced numerous challenges and cuts over the years. Despite its significant role in supporting cultural programs, the NEA's funding has not kept pace with inflation and increasing demand. This trend is mirrored globally, where economic pressures often lead to reduced government support for the arts. In the United Kingdom, for instance, Arts Council England has also experienced big funding reductions, impacting the ability to sustain cultural projects and institutions.[24]

In a discussion with our team at Future of Cities®, Sunil Iyengar, the director of research and analysis at the NEA, had this to say: "In general, I don't think people recognize how much of an economic footprint the arts have in this country. We are for sure seeing some setbacks for major industries in the arts, but we're seeing resilience too."[25] The stats agree. The most recent Americans for the Arts report highlights that, for every dollar spent on arts and culture, there is typically a US$7 economic return in the form of job creation, tourism, and community development.[26]

As regenerative placemakers, we must prioritize arts and culture and must be willing to lead a movement to protect them. Our ventures in Florida, especially with the Wynwood Arts District and the Phoenix Arts and Innovation District, are more than just projects; they are vibrant celebrations of arts and culture that are paying off in ways we couldn't have imagined.

Art in Action: Wynwood Arts District

The Wynwood Arts District in Miami, Florida, stands as one of the biggest success stories in the integration of arts and culture into a place. Wynwood has evolved into a cultural and creative epicenter, unparalleled in Miami, and is perhaps the most famous street art district in the world. It achieved a lightning-in-a-bottle moment when Instagram was born just as muralists and street artists were transforming the area with vibrant, eye-catching murals. Visitors began flocking to the district, snapping selfies and sharing the art with the world via the social media site, accelerating its rise to global fame.

This transformation is a powerful example of how art can revitalize and uplift communities, and it has created a homegrown, exportable brand. Local businesses like Zak the Baker and Panther Coffee proudly use the moniker "Made in Wynwood" to emphasize local pride, similar to how Brooklyn has cultivated its own cultural marketing appeal. Wynwood's success shows that when you get culture right, it not only strengthens local identity but becomes a powerful, exportable brand that resonates far beyond the place's borders.

This beautifully colorful neighborhood, once a neglected industrial area, has become a global destination for creativity thanks to its mural art. This transformation has not only created a unique cultural identity but also wildly boosted the economy through additional local businesses and tourism. Its transformation was thanks to visionary developers. Among them was Tony Goldman, one of the masterminds behind the revitalized SoHo and South Beach's Art Deco Historic District, but he was not alone. A group of us, including artists and creatives, built Wynwood from the ground up.

My journey in Wynwood began in 2000, right after getting my

real estate license, when I started working with Nina Arias, the artist and businesswoman I introduced you to at the beginning of the book. She cofounded the Wynwood Arts District Association. By 2005, Goldman was looking at properties in the neighborhood. We had seen the opportunity early on, and by then, the Wynwood Arts District Association had already been formed, setting the stage for the neighborhood's incredible transformation.

Tony Goldman, alongside his children, Jessica and Joey Goldman, played a pivotal role in shaping Wynwood, contributing to iconic projects like Wynwood Walls and championing initiatives like the Neighborhood Revitalization District rezoning, which transformed the neighborhood. Other influential pioneers included Mera and Don Rubell, whose significant support of the local art scene left an enduring impact, as well as renowned collectors Marty Margulies and Constance Collins. Constance, notably, founded Lotus House, a Miami shelter for homeless women facing abuse.

The area's artistic growth was further fueled by early players like Fred Snitzer, one of Wynwood's first gallerists, and Primary Projects, led by Books Bischof and Typoe, who nurtured street art talent and fostered cultural innovation. Developer David Lombardi introduced residential life to the district with Wynwood Lofts, the area's first residential building, while key broker Larry Mizrach, along with my team and me at Metro 1 Commercial, helped secure the vital warehouse deals that laid the groundwork for Wynwood's emergence as a vibrant arts hub.

At first glance, however, you probably would have called us optimistic. There were so many warehouses, so many dull, blank walls, and not a lot of culture to be detected by the untrained eye. Deciding to help develop Wynwood, we drew on regenerative frameworks, such as permaculture principles, placemaking, and biophilic design. We knew that, if nothing else, art and culture would be the keys to bringing Wynwood out of the shadows and into the limelight. And they were.

As William D. Talbert, president and CEO of the Greater Miami Convention & Visitors Bureau, said, "There would be no Wynwood without Wynwood Walls. Where I saw warehouses, [they] saw an outdoor art museum. That's the difference between a visionary and the average Joe."[27]

We kicked off the project by collaborating with a community of artists, among them, Eduardo Kobra, one of my favorites, to identify spaces that could support their work. Their creativity was a vital resource, and they were treated as the experts when it came to bringing the walls to life. Never once did we developers dictate what should and shouldn't be painted. The artists had both the skills and the cultural awareness to make it happen.

One of my favorite murals in Wynwood is a stunning piece by the remarkable Lady Pink. As a pioneering street artist, Lady Pink was intent on honoring the cultural diversity and resilience of the Wynwood community. The mural depicts a reclining woman whose body is made up of pink bricks, surrounded by vibrant plant life that reflects Florida's lush nature. This artwork is significant for its celebration of regeneration and the theme of rebuilding a sense of home. It also highlights the historically overlooked contributions of women, as well as Latinos and immigrant cultures, that have shaped Miami. Lady Pink's wall, as well as the other hundreds of walls of Wynwood, now attract around ten million visitors per year looking to get a glimpse of Wynwood's culture. A book was even written about its story. If you'd like to read more, check out *Walls of Change: The Story of the Wynwood Walls* by Jessica Goldman Srebnick and Hal Rubenstein.

Simultaneously, as the artists were working their magic, we began advocating for the interests of local business owners and proprietors. By hosting some of the first meetings of the Wynwood Business Improvement District in our office and with me becoming one of the founding board members, Metro 1 and I helped costeward the revitalization efforts that would shape Wynwood's future.

One of our big wins was advocating for progressive zoning reforms (the Wynwood Neighborhood Revitalization District), which allowed for higher-density developments while preserving the area's cultural character. These zoning changes were instrumental in attracting investment and generating funding to redesign a place where local artisans, business owners, and street artists could flourish together. From art galleries and live music venues to cultural restaurants and market events, local enterprises have thrived, contributing to the neighborhood's economic resilience alongside its cultural identity. As Sunil Iyengar advocated, art can equal financial abundance for a place that far surpasses expectations. Art is a win–win–win for community, culture, and the economy. The district's transformation from a neglected area to an artistic capital shows just how art can revitalize and celebrate a previously overlooked culture.

From Wynwood, artistic effects have rippled out to the rest of the state. Art Basel, the renowned international art fair, made its Miami Beach debut in 2002, thanks in large part to the efforts of Mera and

Don Rubell, who had one of the largest private art collections in Miami located in, you guessed it, Wynwood.

That art fair then expanded into Miami Art Week, now featuring dozens of satellite fairs all over the city, from Wynwood to the Design District, and drawing more private jet traffic and art lovers in December than any other event in the state. It brings in around US$500 million per year to the economy.

Craig Robins, the visionary founder of the luxury Design District, further fueled this transformation by creating Design Miami, a design and furniture fair that was initially based in the Design District adjacent to Wynwood. Robins's efforts were also a catalyst for the growth and development of the area, helping to solidify Miami as a world-class destination for art, design, and culture. Thanks to these artistic initiatives, Miami has shaken off its old Scarface and Miami Vice image and reinvented itself as a cultural powerhouse. It's less "say hello to my little friend" and more "say hello to my collection of world-class art."

Celebrating Homegrown Culture: PHX-JAX

In the Phoenix Arts and Innovation District, in Springfield, Jacksonville, we've been engaging community members and planning green initiatives, such as community food forests and bike lanes, which will be part of the future Emerald Trail project. What's key to our approach, however, is cultural preservation. Throughout the entire duration of the project (that's still ongoing), we've been building upon a preexisting culture that was homegrown and organic. The Phoenix Arts and Innovation District had an artistic identity long before we became involved. Rather than impose a new direction, we've embraced and honored the existing culture, working to amplify what was already thriving here. By building upon this strong local foundation, we've been able to create a project that not only respects the current culture but enhances it, giving it the attention it deserves.

Take PorchFest, for example, a beloved annual event that was already up and running in the area before we got there. The locals loved it, but at the time, it was a special event that needed support to showcase the neighborhood and draw in more attendees. So to support our regenerative efforts alongside PHX-JAX, we decided to support and sponsor this local celebration, which is all about community, historic preservation, and the charm of front porches—an architectural feature that is becoming largely lost in modern America. It now draws in over 10,000 people every year. PorchFest brings the neighborhood to life with live music performances on porches and local food vendors and art displays that line the streets. It's a day filled with music, laughter, and togetherness, where families, friends, and visitors stroll from porch to porch, enjoying performances from local and regional artists. It's amazing, and if you haven't attended yet, you should!

This event is a perfect example of how festivals can transform neighborhoods by bringing people together in the name of culture. Rather than creating something new, we approached this preexisting festival with respect and care and sought to amplify it. It's a testament to how my team and I approach regenerative placemaking in general: taking what's already working in the community and giving it the platform and support it needs to flourish.

A city is nothing without its community, and a community's glue is its culture. It's also incredibly challenging to get community members on board with a project without showing your interest and commitment to integrating their culture and their art into your plans. A big part of doing that in PHX-JAX was to showcase local artists and their work as an integral theme of the area, just like in Wynwood. Although this was on a much smaller scale, and the project is still in its growing phase, the area is now teeming with local creatives, and the walls are painted with some of the most stunning street art in the city.

Take the Emerald Woman, for example, a ginormous mural painted on our Emerald Station community event center and workspace. Adorned with an emerald jewel on her forehead and a leafy crown, she represents the neighborhood's involvement in the city's famous Emerald Trail, as well as the neighborhood's vision of being home to a city of the future that thrives alongside nature. She is a symbol of pride for all who reside there. Among countless pieces of incredibly vibrant and inspiring street art, there is an amazing mural of a young African American boy named Kalif, painted by his father. Due to popular demand, our artist Chris Clark went on to paint a series of portraits featuring this local arts family in Jacksonville. Many local artists from the area have found more artistic opportunities than ever before through PHX-JAX. Kalif was painted during the Spring Mural Jam, a public event where fifty local artists transformed the area into a living art museum in real time.

Then there was the 48 Hour Mural Festival, which turned 14th Street into a dynamic open-air gallery. The murals were highly relevant culturally, such as those by Toneism, Trevor Locy, and Boob

Boom Franzi, who paid tribute to local hip-hop legend Paten Locke, celebrating Jacksonville's artistic heritage and contemporary creativity.

The Phoenix Holiday Art Fair and Makers Market came afterward, the wonderful event that transformed a former industrial warehouse into a bustling marketplace to showcase and sell local produce in December 2023. Organized by our local resident powerhouse women of Future of Cities® Tanya Watts, Emily Moody, and Suzanne Pickett, the market celebrated Jacksonville's diverse cultural scene, provided vital support to local growing businesses, and helped strengthen community ties, despite some really challenging weather conditions! The collaboration of these women and all the teams involved in the regeneration of PHX-JAX has been instrumental in this process, leveraging their grassroots connections to build genuine community relationships and ensure the project resonates with local values.

None of these cultural events, and more, are one-time occurrences. They are now happening on a regular basis to keep up cultural momentum. Just as culture never sleeps or ceases to evolve, neither does the flow of our events. Our focus on celebrating and preserving local culture has been the driving force behind reclaiming a once-forgotten local identity. By integrating arts and cultural programming deliberately and strategically, we not only honor the area's history but also enhance its appeal to investors and sponsors.

We're proud to share that our commitment to cultural development has been formally recognized by the mayor of Jacksonville and the Jacksonville City Council, who have unanimously approved our designation as the planned developer and awarded US$5.5 million in grants.

The City of Jacksonville was also awarded a historic US$147 million grant from the Department of Transportation in March 2024. This is the largest-ever grant awarded to Jacksonville. It will completely fund the most important piece of regenerative infrastructure in the city, which will help transform the Phoenix Arts and Innovation District into a thriving hub of culture and innovation. Protecting, preserving, and showcasing culture is no simple task; it's a nuanced process that requires time, patience, and a deep commitment from regenerative placemakers. It demands active listening, empathy, and an understanding of the community's unique needs based on their cultures. Sparks can fly if you don't approach this process with care and awareness—challenges we've encountered personally since founding Future of Cities® in the year 2000. Speaking from our experience, let's delve into some of those challenges.

Cultural Complexities 101

I'd love to be able to say that a place's culture is delivered and presented to placemakers in a neat package, that a designated and friendly cultural ambassador will sit you down and clearly outline what a place's culture is and how you can leverage it in your regenerative placemaking project. But this beautiful fantasy has never come to fruition for me, my team, or anyone else I know who has ever worked on regenerating a diverse place that's never been touched before. It requires a lot more effort and proactivity than that. That's not to say that cultural ambassadors don't exist or that the community won't support you and be instrumental in your integration of this third pillar. It's just that culture is such a broad and subjective concept that it goes beyond the perception of one voice, two voices, three voices, or even a hundred voices of the area.

Culture operates on local, domestic, and international levels, from the tight-knit nature of a family, street, neighborhood, and small town to the complex dynamics of an urban metropolis, and, ultimately, a country and global identity. And just like it's tough to define in the first place, it's also tough to perfectly protect and enhance. At its core, culture is the shared values, beliefs, customs, and practices that bind a community together. However, these elements do not exist in isolation either. Culture is shaped by the local environment and systems too. Domestic influences, like national identity and historical context, also play a huge role in shaping cultural practices. Finally, international trends and global interactions further complicate the cultural landscape, introducing new ideas and practices into the local mix. When working on regenerative placemaking projects, it's very important to recognize and respect these layers of culture. A one-size-fits-all approach simply doesn't work. Instead, as regenerative placemakers, we face the challenge of understanding the unique cultural dynamics of the communities we are working with and acknowledging the local, domestic, and international influences that shape them.

In our experience, more often than not, there are multiple cultures present within even the smallest community. Think about your own town or city. I doubt there is just one religion, just one race, just one music taste, just one key sport, just one interpretation of history. Therefore, multicultural preservation is key, especially in places that are melting pots of people with origins from all over the world, like New York and London. It's paramount that different community members with varying cultural backgrounds feel recognized and valued within your project, as well as within their communities. As regenerative placemakers, we are pioneers of diversity, bridging the gaps between social and ethnic groups to bring about a sense of mutual respect.

One simple yet effective way to integrate multiple cultures in a regenerative placemaking project is by organizing a community festival that celebrates the diverse cultural traditions present within the area. At the festival, different cultural groups could be invited to showcase their music, food, dance, and art. This event would allow the community members to experience and appreciate each subculture and, ultimately, create a collective culture that's built on not just tolerance but pride. To make an event like this truly inclusive, our advice would be to involve local ambassadors or well-known community leaders from each cultural group in the planning process.

Building culture within regenerative placemaking is not just about creating physical spaces or organizing events. It's about building relationships and trust within the community. This requires empathy, deep listening, and a willingness to engage in difficult conversations.

Cultural tensions can arise from a lack of understanding or from historical grievances that have not been addressed. Regenerative placemakers must be prepared to listen to the concerns and needs of different cultural groups and to facilitate dialogue that, hopefully, leads to healing and understanding. This can be a slow and challenging process, but it is essential for creating a culture of respect.

Solutions for Overcoming Cultural Roadblocks

Despite the best intentions, cultural roadblocks are inevitable in any regenerative placemaking project. These can range from resistance to change within the community to misunderstandings between different cultural groups. In our experience at Future of Cities®, the key to overcoming these challenges is patience, adaptability, and commitment, combined with the following strategies:

Lean on Local Leaders

Do not play a guessing game when it comes to understanding a place's existing culture. Involving local cultural leaders in the planning and implementation process can help exponentially speed up your progress and build trust within the community faster. These leaders can act as cultural ambassadors, helping to translate the goals of the project into culturally relevant terms. Ask the right questions about a place's history and which subcultures are present within it.

Celebrate Every Small Win

Cultural understanding and change are not always immediate; they often unfold gradually, with each small success laying the groundwork for more significant transformations. In the context of regenerative placemaking, celebrating these small wins is essential for sustaining momentum and enjoying a sense of progress. When a community sees even the smallest tangible results—such as a successful cultural event, the creation of a new public art piece, or the launch of a community garden—it not only builds confidence among the participants but also validates the efforts of those involved. These moments of recognition reinforce the value of cultural inclusivity and the broader mission of the project.

Provide Education and Build Awareness

Educational initiatives that raise awareness about the cultural diversity within the community can help reduce prejudice and promote understanding. By offering programs that highlight the richness of cultural diversity, communities can break down barriers and build a stronger, more inclusive environment. One innovative approach is the creation of a cultural guild—a collective of local residents, artists, and educators who come together to share knowledge, preserve traditions, and mentor others. A guild can serve as a hub for cultural exchange, offering workshops, skill-building sessions, and events that engage the community in hands-on learning.

Create Safe Spaces for Open Dialogue

Last but by no means least is the critical practice of listening deeply to community members and residents. Their insights into local culture are invaluable and can profoundly inform your regenerative placemaking project. So, if you are in doubt, just *ask*. Creating forums where community members can share their cultures and engage in open dialogue with each other is essential for building that much-needed base of trust. These spaces should be designed to be inclusive and respectful, ensuring that all cultural groups feel encouraged to contribute. The culture toolkit at the end of this book offers some ideas on the questions you should be asking.

Despite the complexities, the power of culture in regenerative placemaking cannot be overstated. When culture is nurtured and integrated into the fabric of the community, it can transform a place, creating a sense of belonging and pride. It can also serve as a catalyst for economic development, attracting visitors from outside the area and supporting local businesses. Ultimately, the success of regenerative placemaking depends on the ability to navigate the cultural complexities of the communities we serve.

The Interplay of the Three Pillars of Regenerative Placemaking

Culture is deeply intertwined with the natural environment. The natural elements of a location—its plant life, animals, geography, and climate—play a bigger role in shaping its cultural identity than you might think.

For example, the cold, snowy winters of Scandinavia have significantly influenced the cultures of countries like Sweden and Denmark, where the traditions of *hygge* and *fika* were born. The harsh climate has shaped a national appreciation for the warming rituals associated with cozy indoor gatherings by the fire with hot tea, coffee, and chocolate.[28] On the flip side, sweltering countries like Qatar and the United Arab Emirates offer a refreshing escape for locals in air-conditioned cafés nestled in shady, narrow alleyways. Here, locals enjoy chilled yogurt-based drinks and cold desserts, such as the renowned rosewater pudding known as *muhallebi*, a staple in traditional Middle Eastern cooking.[29]

Then you have a place's flora and fauna. For example, in Australia, the eucalyptus tree and its koala inhabitants have shaped both Indigenous and modern cultural expressions. The eucalyptus tree is integral to Aboriginal art and storytelling, while the koala has become a beloved national icon.[30] It's the same with the American bald eagle in the US, which symbolizes freedom and strength and is prominently featured in national emblems and currency. Because of this, the bald eagle has inspired numerous conservation efforts to protect this particular bird over others, and that's a testament to the power of the deep connection between cultural heritage and environmental stewardship.[31]

Another example of the cultural impact flora and fauna can have can be seen all over Japan. Japan's cherry blossoms, or *sakura*, are deeply ingrained in ancient cultural practices and celebrations. The arrival of spring is marked by *hanami* festivals, where people get together to view and celebrate the fleeting beauty of these blossoms, an act of collective reverence for nature's transience and the changing seasons.[32]

As you can see, the natural environment not only influences but often defines the cultural practices, symbols, and identities of a place. Knowing this as regenerative placemakers is useful because we can beautify and diversify places through the introduction of local native plants in green spaces as well as native animals (in art and educational initiatives; of course, I doubt that Jacksonville would have appreciated the introduction of wild mountain lions to the Arts and Innovation District). Murals and sculptures featuring local wildlife can be a nice way to pay tribute to the place's natural heritage.

Understanding and integrating the unique natural characteristics of a place into regenerative placemaking creates very meaningful, calm, and beautiful spaces. And the good news is that very few people on this planet would be against more nature and greenery in their environment. Nature is our ancient home, and although many of us live in concrete jungles now, in a way, it always will be.

You could say that nature is the culture shared between us all. So as we continue to develop and regenerate our urban landscapes, it's important to remember that the natural world is not just a backdrop to but a vital component of our cultural heritage, not to mention a key factor in shaping our collective future.

The interconnection and interdependence among the three pillars—community, nature, and culture—form an unbreakable foundation that defines any place. These pillars are universally applicable across all placemaking projects, yet they uniquely shape the identity and dynamics of each urban experience. Each pillar supports the others, creating a symbiotic relationship where progress in one area fuels success in another. Similarly, we must be careful of instances where success in one area can damage another pillar.

As a basic example, let's say the introduction of a café to a place can boost community and culture but may have a negative impact on the environment depending on the way it's run. Similarly, when we introduce cultural events, we should think about how they can strengthen community bonds and be environmentally sustainable, ensuring that every action we take is mindful of the broader system it influences. It takes innovation and creativity, but it's almost always possible to strike a balance.

This interconnectedness is a key aspect of systems thinking, where we approach urban development holistically. This holistic approach is in line with the philosophy of Bill Reed, a renowned systems thinker and developer. “Regeneration is not simply about making a landscape and local habitat more productive and healthier. Effective regeneration requires that we engage the entirety of what makes a place healthy. This may be our home community, a corporate campus, a small lot, or a building. When we start from a whole-systems understanding, any of these entities is an entry point into the whole system. Each is an integral part of a living system, and a key role can be found for anyone and any system within the smallest to largest physical footprint. The footprint is not the limiting factor, as long as a sense of conscious engagement can be realized by the people who are part of it.”[33]

The success of any place hinges on recognizing that every single person involved, as well as every element of the place, no matter how small, contributes to the larger system. When we cultivate conscious engagement among the people who inhabit these spaces, we chart the way for a truly regenerative and sustainable urban environment, where community, nature, and culture exist in harmonious balance.

CHAPTER 6

Measure What Matters

Not everything that can be counted counts,
and not everything that counts can be counted.

—JOHN DOERR

After integrating the pillars of community, nature, and culture into a regenerative placemaking project, it's important to reflect on how strongly these pillars are standing. It's essential to assess whether the place, its people, and the natural environment have truly improved as a result of our efforts. The first step in this evaluation is to recognize that measurement is part of an *ongoing* developmental process, rather than a method of assigning a fixed score or grade to the effort. You don't just measure success once. You'll measure it again and again and again. Measurement in regenerative placemaking should be seen as a tool for learning and growth, helping the project evolve over time. So, this is useful for when progress seems slow. Don't lose hope. Read the results, adapt, have patience, and read them again. Once you're measuring it, you're managing it.

The second step is to accept that measurements must also adapt and change as the project matures. In regenerative placemaking, both fixed and adaptable system measures play a role, and these can evolve in response to changing conditions and stakeholder input. You want to be asking two key questions: What are the enduring indicators of a healthy and vibrant place (fixed measures)? And what are the unique, evolving traits of the place that signal its growth and abundance over time (adaptable measures)?

Although the specific metrics and benchmarks will shift as the project progresses, the overarching goal remains constant. Regenerative placemaking aims to enhance social and ecological outcomes for everyone. This means assessing the richness and quality of relationships—whether between individuals, within the community, or between the built and natural environments—that contribute to these positive outcomes. By focusing on these interconnected relationships, we can truly gauge the success of a regenerative placemaking project.

The PPP Framework (with an Extra P?)

The PPP framework—people, planet, profit—serves as an invaluable tool for assessing holistic success, not only in regenerative placemaking but across many industries. In fact, most truly successful businesses use it to measure progress in a more well-rounded way. While comparing regenerative placemaking—with its deep-rooted values and spiritual undertones—to the world of business might seem unconventional at first, the two share significant common ground, and the PPP framework is one key connection. Both aim to create systems that are sustainable, resilient, and capable of evolving over time. A successful business relies on strong, reciprocal relationships with its stakeholders, and regenerative placemaking thrives on similar connections. In this way, the two worlds are more aligned than you might think.

Also known as the *triple bottom line*, the PPP framework is a well-known sustainability model that evaluates success in three key areas. Traditionally, businesses focused solely on financial returns, or the profit aspect. However, there is a growing recognition of the need to consider environmental and social returns—people and planet—as equally important. This shift is perfectly aligned with the principles of regenerative placemaking, where the goal is to redesign cities and neighborhoods that are not just economically successful but also socially equitable and environmentally sustainable.

The PPP framework was introduced in 1994 by John Elkington, a pioneering business thinker and founder of SustainAbility. At the time, it was a revolutionary concept, encouraging businesses to adopt a more holistic approach by shining a light on social and environmental concerns alongside profits. Rather than focusing on a singular bottom line, the triple bottom line approach values people, planet, and profit equally.

However, in regenerative placemaking, there's often an additional metric that comes into play: purpose. This fourth P is just as crucial as the others, although it is more difficult to quantify. Purpose ensures that every action taken in a project is aligned with the higher mission of improving life for all—whether it's human communities, ecosystems, or the cultural heritage of a place. At every step, you should ask, *Is this in line with our purpose of making this place better for all life connected to it?*

With the addition of purpose, you could call this the quadruple bottom line. While profit, people, and planet are measurable to a degree, purpose acts as the guiding principle, ensuring that every decision is rooted in long-term, meaningful change. Purpose might not come with precise metrics, but it serves as the North Star, asking whether the project enhances holistic well-being for everyone and everything involved.

In regenerative placemaking, while the pillars—community, nature, and culture—may differ slightly, the quadruple bottom line offers a well-rounded framework. It integrates not just financial and environmental outcomes but also the deeper vision of improving and supporting all life through purpose. After all, balancing purpose with people, planet, and profit is what makes placemaking truly regenerative and enduring.

Let's dive deeper into the PPP framework and explore how it can be applied to measure success in regenerative placemaking.

People

The people component is focused on the social outcomes of the project, including how it affects community well-being, equity, and inclusivity. Measuring success in this area involves assessing the benefits and opportunities provided to individuals and communities, ensuring that the project enhances the quality of life of the community. The relevant metrics might include surveys on resident satisfaction, the level of community engagement, intercommunity relationships, the frequency of cultural events, and improvements in access to services and opportunities. The people component is covered by the pillars of community and culture in our regenerative placemaking projects.

Planet

The planet component is the environmental impact of the project, evaluated by focusing on cleaner sustainability practices and their outcomes. The key metrics include resource conservation, such as reductions in energy consumption and water use; waste reduction, including efforts to minimize waste and enhance recycling or upcycling; and ecological restoration, which measures improvements in natural habitats. Success is determined by how well the project reduces pollution, decreases food waste, increases green spaces, builds resilience, and enhances biodiversity, ultimately contributing to the overall health of the natural environment of a place. It's about ensuring that the placemaking efforts do not deplete or harm the environment but rather contribute to its regeneration. The planet component is covered in the pillar of nature in our regenerative placemaking projects.

Profit

The economic aspect assesses the project's financial viability and benefits by evaluating job creation, local economic growth, and overall financial sustainability. The key metrics include revenue generation, cost efficiency, job creation, business expansion, grants, and other economic contributions to the area, as well as the return on investment and the broader economic impact on the community. Analyzing these financial data points provides clear indicators of economic success and challenges. In regenerative placemaking, however, profit extends beyond traditional financial returns; it involves many types of capital, including social and cultural wealth. This approach prioritizes supporting local small businesses, creatives, and artists as a powerful way to revitalize the community by keeping investments localized, reducing supply chain pressures, and minimizing the carbon footprint associated with globalized needs. Inspired by frameworks like Local Futures by Helena Norberg-Hodge, regenerative placemaking emphasizes local resilience and sustainability.[1] For example, government grants could be assessed not only for their financial return but also for their impact on community well-being, cultural events, and renewable energy initiatives. Another example might be the financial success of a new local café that serves as a cultural hub, community gathering spot, and model for zero-waste practices, embodying all three pillars of regenerative placemaking.

Incorporating the PPP framework into regenerative placemaking evaluations provides a more holistic understanding of a project's impact. Regular research and reporting should be conducted throughout the project and should continue annually after its completion. Be ready to adapt to evolving initial findings and shifts in key objectives. For example, initially, we might focus on metrics like revenue and job creation to gauge immediate financial results. Over time, as the project matures, we shift to metrics that reflect long-term impacts, such as business growth, market share, and overall economic resilience, which takes years. Have patience. This adaptability keeps the evaluation process relevant and helps us capture both early achievements and ongoing benefits, ensuring a clear view of the project's economic performance and its contribution to the local community.

Remember: While the measures can shift, the objective stays the same: to create sustainable cities of the future where community, nature, and culture thrive.

Case Study: The Cheonggyecheon Stream Restoration Project

A great example of the intersection of community, nature, and culture (although not a full representation of regenerative placemaking's potential) is the Cheonggyecheon Stream Restoration Project. The Cheonggyecheon Stream Restoration Project was an urban renewal initiative in Seoul, South Korea, that transformed a long-covered, polluted stream into a vibrant, public green space. Initially buried under concrete during the twentieth century to make way for roads and an elevated highway, the stream was uncovered and restored between 2003 and 2005. The project aimed to serve the pillar of nature by reducing urban heat, increasing biodiversity, and providing flood protection. It also aimed to enhance the pillars of community and culture by creating a cultural and recreational area that attracted millions of visitors, boosted local businesses, and revitalized the quality of life for surrounding neighborhoods.[2]

Here are some examples of how they're measuring it within the PPP framework:

People (Community and Culture)

The project attracts an average of 64,000 visitors daily, including 1,408 foreign tourists, who contribute up to 2.1 billion won (US$1.9 million) in visitor spending to the Seoul economy. It improves health by lowering respiratory disease rates in the area among community members.[3] It significantly boosted community satisfaction; the residents now enjoy a revitalized natural space for recreation. The clean river has become a popular spot for activities like bathing and fishing, contributing to a higher quality of life and a deeper connection with nature for the people of Seoul.[4] And it contributed to a 15.1% increase in bus ridership and a 3.3% increase in subway ridership in Seoul between 2003 and 2008, enhancing quality of life and convenience.

Planet (Nature)

The project provides flood protection for up to a 200-year flood event, designed to handle rainfall intensities of up to 118 millimeters per hour. It also increased overall biodiversity by 639% between 2003 and 2008, with significant growth in plant, fish, bird, insect, mammal, and amphibian species, with plant species increasing from 62 to 308, fish species from 4 to 25, bird species from 6 to 36, aquatic invertebrate species from 5 to 53, insect species from 15 to 192, mammal species from 2 to 4, and amphibian species from 4 to 8.[5] It reduces the urban heat island effect, with temperatures along the stream 3.3 degrees Celsius to 5.9 degrees Celsius cooler than a parallel road just blocks away. This cooling results from the removal of the paved expressway, the presence of the stream, increased vegetation, reduced auto trips, and a

2.2%–7.8% increase in wind speeds through the corridor. It also reduces small-particle air pollution by 35%, lowering respiratory ailment rates in the area.

Profit

The project increased land prices by 30%–50% for properties within fifty meters of the restoration project, double the rate of property increases in other areas of Seoul. That led to more businesses; it increased the number of businesses by 3.5% in the Cheonggye-cheon area during 2002–2003, double the rate of business growth in downtown Seoul. It increased the number of jobs available in the area—an increase of 0.8% workers over the first year of implementation, compared with downtown Seoul, which saw a decrease of 2.6%. Finally, the project received the equivalent of US$280 million from the South Korean government.[6]

As you can see, many of the measurements overlap within a PPP framework, and a win in one category can be a win in another as a happy accident. This project also demonstrates how vital it is for any element flowing through a community—such as water or money—to contribute positively to the area. For instance, water flowing through a community is of little benefit when confined to a pipe, where it cannot enhance habitats, provide urban cooling, or support recreation. Similarly, money only benefits a community when it nourishes and contributes to the local economy. As Michael Shuman has shown, every dollar spent at a locally owned business generates two to four times the amount of jobs, income, wealth, and taxes as a dollar spent at a comparable nonlocal business.[7] The benefits of investing in and valuing regenerative placemaking in this sense are undeniable, as is evidenced by this simple yet very effective stream restoration project. This project highlights the environmental, social, and economic impacts that thoughtful, relationship-focused urban development can have on a wide range of stakeholders.

It's worth noting, however, that no project is perfect. In the case of the Cheonggyecheon Restoration Project, there's been some critique too. Critics argue that all these positive outcomes came at the cost of cultural heritage, with long-standing sites and communities displaced to make way for redevelopment. Environmentalists have also pointed out that the stream's artificial concrete bed lacks natural water purification, leading to water quality issues and algae growth. Furthermore, allegations of corruption and claims of political motivations added to the controversy, with some seeing the project as a vanity move aimed at enhancing Mayor Lee Myung-bak's political profile.[8]

Despite these downsides, Cheonggyecheon's popularity and influence highlight its impactful legacy—serving as a model for urban revitalization worldwide and offering valuable insights into the considerations of transformative and regenerative design.

Case Study: Addis Ababa's Upcycled Housing

In the bustling heart of Addis Ababa, Ethiopia, Kidus Asfaw found himself surrounded not only by unfinished building projects but also by mountains of plastic waste in the city. This pollution wasn't just an eyesore; it was a crisis fueling huge-scale environmental degradation, and it was hindering local development. Driven by a vision to transform these challenges into opportunities, Asfaw cofounded the startup Kubik in 2021 with an ambitious goal: to transform this plastic waste into a better future.

Kubik is using plastic waste to create something extremely valuable: homes. The company has developed an innovative process that converts hard-to-recycle plastic into low-carbon, cost-effective building materials. These materials are not just ecofriendly; they are also affordable, providing a much-needed solution to the severe housing shortages in rapidly urbanizing African cities like Addis Ababa. With each modular block produced, Kubik diverts tons of waste from landfills and oceans, turning environmental crises into opportunities for sustainable urban development.

With forty-four million Ethiopians, or 33% of the population, living in extreme poverty, Kubik's operations are deeply rooted in regenerative missions to create clean and affordable living for all while significantly reducing the environmental footprint of construction.[9] Here are some examples of measurable progress across the three pillars of regenerative placemaking with a PPP approach:

People (Community and Culture)

Kubik provides cost-effective housing solutions that are significantly cheaper than traditional builds, addressing housing shortages in rapidly urbanizing cities like Addis Ababa. This project reduced construction prices by 40% (when compared with traditional cement-based construction).[10] The manufacturing process of Kubik's building materials has created hundreds of employment opportunities in Ethiopia, contributing to economic growth in local communities. As they scale across Africa, thousands more jobs will become available to people who need them. By providing safe, dignified, and affordable housing, Kubik is improving the quality of life for vulnerable populations. This promotes social equity. Additionally, Kubik is actively working to improve conditions for waste collectors, offering safer working environments and stable incomes.[11]

Planet (Nature)

Kubik's production process generates up to five times less carbon dioxide than traditional construction methods. Their low-carbon building materials produce 80% less carbon emissions than conventional cement.[12] Kubik initiatives divert tons of plastic waste from landfills and oceans, approximately forty-five tons of hard-to-recycle plastic waste per day. Plastic recycling on this scale helps protect marine life and other wildlife from the devastating effects of plastic waste that ends up polluting natural habitats. The modular building blocks allow for the easy assembly and disassembly of structures, making them suitable for disaster relief efforts and areas with rapid population growth. Kubik's approach allows construction to be completed two to three times faster than traditional methods, requiring low-skill labor.

Profit

Kubik's operations have attracted over US$5.2 million in seed funding to scale production and expand across Africa. Kubik is planning to expand into other African countries, leveraging its recent funding round and motivating new investors. Kubik has secured significant partnerships and funding from entities like the Global Startup Awards and International Finance Corporation that have a special interest in the success of the project.[13] By converting discarded plastic waste into building materials, Kubik derives economic value from "worthless" waste, reduces the need for new raw materials, and supports sustainable economic growth.

Kubik's journey is still in its early stages, but the potential is undeniable. With their innovative approach and unwavering commitment to sustainability, Kubik is not just building homes; it's building a better future for people and our planet. Kubik's innovative approach to transforming plastic waste into affordable, low-carbon building materials exemplifies how sustainable practices can address urgent urban challenges while promoting economic growth and social equity.

However, no regenerative project is perfect, and Kubik's model presents certain potential drawbacks. For one, using plastic in construction may pose health and environmental risks, because certain plastics can release harmful chemicals when exposed to extreme temperatures, potentially impacting the residents and nearby ecosystems.[14] Additionally, while Kubik's low-carbon materials are ecofriendly, their durability compared with traditional options like concrete is still under examination. If these materials prove less durable, buildings may require more frequent repairs, which could offset some of the initial environmental benefits.[15]

Stakeholders and Their Influence over Your Metrics

Measuring the impact of a regenerative placemaking project can involve a wide range of metrics, which can feel overwhelming. That's why, at Future of Cities®, we've developed a flexible framework to help guide your evaluation process, which I will share with you at the end of this book. Before you dive in, it's crucial to ask yourself, *Who are our stakeholders?* The answer to this question will significantly shape the metrics you choose to measure success. Identifying your stakeholders early on will ensure that your project aligns with the needs, concerns, and expectations of those it impacts, guiding you toward more relevant and meaningful metrics.

For anyone unfamiliar with the term, a *stakeholder* is anyone who has an interest in or is affected by the outcomes of your project. This could include individuals, nested wholes, groups, or organizations. Stakeholders can be involved in different ways, either by influencing the project, being directly impacted by it, or having a stake in its success or failure. For this reason, stakeholders can also be plants and animals, as well as entire ecosystems. Within the framework of people, planet, and profit, the stakeholders for a regenerative placemaking project might include the following:

- Local residents
- Community members
- Community leaders
- Community centers
- Local businesses
- Government agencies
- Nonprofit organizations

- Environmental agencies
- Local ecosystems
- Conservation groups
- Utility providers
- Climate advocates
- Investors
- Developers
- Real estate agents
- Suppliers
- Natural stakeholders, such as wildlife, waters, soil, and plant life

Stakeholders are important because their support, opposition, or participation can significantly influence the success of a project and, moreover, the definition of that success. Engaging stakeholders effectively means considering their needs, concerns, and contributions and ensuring they are part of the decision-making process whenever possible, from the very beginning.

Next, consider establishing or supporting existing guilds—networks of interdependent, mutually supportive relationships—among your stakeholders. These guilds strengthen the resilience and sustainability of your project by fostering collaboration, where each member contributes to and benefits from the collective effort. By organizing or cocreating guilds, you shift the focus from isolated components, like individual buildings or community groups, to larger, interconnected systems—whether urban ecosystems, regional economies, or expansive social networks.

A guild can be the foundation of a community or project. For example, in a regenerative placemaking project, the local residents might come together to form a guild around a community garden. This garden isn't just a stand-alone effort; it's part of a bigger picture—a nested system that includes the local ecosystem (soil, plants, and wildlife), social connections (community engagement and cultural practices), and economy (food production and small businesses).

In this guild, the residents might volunteer for planting, businesses could sponsor supplies, and conservation groups might offer guidance on sustainable practices. Even the plants, insects, and birds in the garden play a role, making them stakeholders in this interconnected system. This approach encourages us to map our projects within the wider systems that affect the cities and regions they inhabit, transforming relationships from merely transactional to truly reciprocal. These reciprocal relationships create mutually beneficial processes and outcomes, not just within the project but across various connections—between the project and its place, between local and global businesses, and among city residents from different districts, cultural groups, artistic scenes, natural ecosystems, and beyond.

Another good example of a guild of stakeholders could be the residents, developers, investors, local government agencies, and nonprofit organizations that are involved in the Kubik housing project in Ethiopia. They're collaborating to ensure that the project meets affordable housing goals, adheres to environmental regulations, and brings about community engagement. By working together, these stakeholders are achieving outcomes that are more sustainable and beneficial for all involved than what any single stakeholder could achieve alone.

In general, the long-term impact of your project will be shaped by the ongoing, adaptive relationships you cocreate with your stakeholders and the broader systems in which your project is embedded. This relationship with your stakeholders is known as *coevolving mutualism*, which means that these relationships are symbiotic and mutually beneficial, the type that stands the test of time. Start now by identifying your stakeholders and understanding the systems you seek to benefit, and your metrics will become more obvious. From there, you can select the most relevant metrics and create a framework that ensures your project continues to generate value long after its initial goals have been met.

CHAPTER 7

The Bridge Between Ideas and Reality

A vision without action is just a dream.
Action without vision just passes the time.
Vision with action can change the world.

—NELSON MANDELA

Welcome to the part of the book where, initially, your eyes might glaze over. When people think of policy, they oftentimes think of two key words: *boring* and *overwhelming*. But it doesn't have to be. Here at Future of Cities*, we love policy, and we've seen firsthand the impact that implementing great policy can have on a place.

Not everyone reading this book is a government official, community leader, or developer, and therefore, having a grasp on what the concept entails is important. Admittedly, like many other creative people I know, when I first entered into the world of regenerative placemaking, writing policy was something I sought to avoid. It was all written in a language I found difficult to understand, and, frankly, I wanted to be working on "more important" things. That was before I truly understood what it meant and the magic it can facilitate.

What Is Policy?

Policy is the bridge that turns ideas into reality, and whether we know it or not, policy permeates everything—from the way we navigate our streets to how we build our homes and protect our natural environments. But it goes deeper than that. Policies shape nearly every single action we take and dictate the rhythms of our daily lives.

Every single business on the planet has policies, and if you work for one, you're adhering to their policies. If you're in a monogamous relationship, you're adhering to policies. If you have children, you're adhering to policies. Policies are the invisible rules that weave together the structure of our personal, professional, and communal lives, guiding our actions and interactions in countless, often unnoticed, ways.

At its core, policy is a set of guidelines and principles that are crafted to influence and determine what decisions get made. Policy is the blueprint that shapes how *everything* operates—whether it's at the level of a city, a neighborhood, or even within a small family. Of course, for the interests of this book, we'll be exploring policy through the lens of regenerative placemaking, or, rather, exploring regenerative placemaking through the lens of policy. In this context, policy serves as your instruction manual on how to keep a neighborhood or city running smoothly, in a way that serves everybody's best interests.

Why Policy Matters in Regenerative Placemaking

In the context of regenerative placemaking, policy is the backbone that supports all the visible and tangible outcomes we care about—community, nature, and culture. It's what makes sure that, when you design a public park, for example, or create affordable housing or develop a transportation system, these initiatives don't just look good on paper but actually *work* over the long term.

Policy is what helps us answer really important questions like these:

- How do we ensure that affordable housing is truly accessible to those who need it?
- What steps can we take to make public transportation not just available, but efficient, clean, and inclusive?
- How can we design public spaces that are safe, welcoming, and reflective of the diverse cultures within the community?
- How can we protect the local flora and fauna while developing an area?

These are all vital questions we've used in our team meetings at Future of Cities®, and they're conversations everyone can get involved in. You don't have to be a professional policy writer to get the creative, solution-oriented results flowing. Policy might not be as sexy as breaking ground on a new neighborhood festival, getting the biggest town grant in history, or painting the most beautiful murals in the country, but it's the magic that makes those things possible.

Great policy transforms visionary ideas into realities. For example, let's say you, and the community of a place, share a vision for becoming more environmentally sustainable. Without the right policies—like incentives for renewable energy, regulations on waste management, or zoning laws that encourage green spaces—that vision remains just that: a vision. It's a nice idea that doesn't exist in the real world. Policy is what takes that nice idea and breaks it down into actionable steps that can be implemented, monitored, and refined over time. And, through experience, that marks the difference between a community that just survives and one that thrives.

Policy Isn't Just for Governments

Governments play a crucial role in crafting and enforcing policies. But they're by no means the only players in the game. Businesses, nonprofits, community organizations, and individual citizens have a role to play in shaping and advocating for policies that align with a place's values and goals. In fact, some of the most effective policies I've seen emerged from grassroots efforts, where local communities identified a need, rallied together, and pushed for change.

An example of a huge, global policy that started from a grassroots level is Complete Streets policies. Complete Streets policies make up a transformative approach to urban planning that prioritizes the safety, accessibility, and convenience of all users of the roadway, including pedestrians, cyclists, public transit riders, and motorists. Unlike traditional road design, which is often focused primarily on making things comfortable for cars, Complete Streets policies aim to create streets that are safe and accessible for *everyone*, regardless of age, ability, or mode of transportation. These policies are now implemented all over the world, but it all began with passionate community members in Portland, Oregon, and Boulder, Colorado, in the early 1990s.

These community members were fed up with high rates of traffic accidents involving pedestrians, a lack of safe routes for children walking to school, and inadequate crossings and infrastructure for cyclists. So they created new policies to do something about it. These new policies spread like wildfire due to popular demand, and the quality of life in these cities skyrocketed, influencing everything from public health and safety to economics and environmental sustainability.[1]

Policies that are crafted in a vacuum, away from the community of a place, often miss the mark, because they fail to account for the needs and aspirations of the people they're meant to serve. This is why involving a diverse range of stakeholders—residents, businesses, and community organizations—is so critical. No matter who you are, know that you have a seat at this table.

That said, of course, governments still play an important role in making sure that these policies are set in stone—at least in the eyes of the law and at least for now. Governments—local, regional, national, and international like the UN and Paris Climate Accord—have an important role to play in ensuring that policies are in place that set a solid foundation for your placemaking projects and that nobody gets in the way of your progress. They're the ones to contact if there's an existing policy that just isn't working or serving the community. They're also the ones you need to get on board with your new policies, of which there will be many if you do a thorough job of regeneratively evolving a place.

Before we move on to the areas of policy you can explore in your projects and the new policies you can cocreate, here are the three areas of expertise of the government that can help get your project off the ground.

Policy

Although we've just established that there is a wide range of stakeholders that contribute to policy creation, the government holds the keys to making sure it happens. From land use, zoning, development, and small business support to affordable housing and incentives, local governments can be great allies for any urban center or built environment. For example, Wynwood's progressive zoning changes allowed for higher-density developments and mixed-use spaces that made the cultural hotspot what it is today. Try not to see the government as an opponent when it comes to implementing your policies but, rather, as an ally.

Funding and Investment

Governments not only help implement policies but can also provide the funding to back them. For example, as I mentioned in previous chapters, the City of Jacksonville received a massive US$147 million grant from the Department of Transportation to support the environmental and cultural efforts of the Emerald Trail. Additionally, PHX-JAX was awarded US$5.5 million for adaptively repurposing warehouses for the use of small local businesses and agreed to provide office space for Jacksonville Small and Emerging Businesses, the small business incubator for the City of Jacksonville. These significant investments show that when governments believe in your vision and your commitment to the pillars of community, nature, and culture, the possibilities for support and growth are limitless.

Planning Permission

Planning permission isn't just a pain in your neck. (Although I won't sugarcoat it; it certainly can be!) It plays a vital role in shaping urban environments to ensure things like cultural preservation and environmental protection. Without having to apply for planning permission, spaces like national parks would be dotted with hotels, Starbucks, and McDonald's, and nothing would stand in the way of precious heritage sites being knocked down to make room for apartment blocks.

Through the planning permission process, governments evaluate and approve development proposals to ensure they align with community goals and standards. This involves working closely with us as placemakers, as well as with the architects, urban designers, and local stakeholders on our team, to design public spaces that are inclusive, clean, and aesthetically pleasing.

If you'd like to know more about government funding and how to get it, you can explore various resources and agencies that specialize in securing funding for regenerative placemaking projects. Websites like Grants.gov and arts.gov offer grant opportunities that can be specifically tailored for community development and creative and regenerative placemaking. International programs like the European Union's URBACT and organizations like the Urban Land Institute also provide valuable support and funding for sustainable urban projects. These resources can help you navigate the often complex process of obtaining government funding, ensuring that your project has the financial backing it needs to succeed.

If you're interested in equipping yourself with the know-how behind planning permission, I recommend checking out resources like your local government's planning department website, which often provides detailed guidelines on the application process, zoning regulations, and design standards. Additionally, organizations such as the American Planning Association or the Royal Town Planning Institute offer valuable insights, educational materials, and courses on navigating planning permissions. These resources can help you understand the requirements, streamline your applications, and increase the likelihood of obtaining the necessary approvals for your regenerative placemaking project.

Policy as a Living, Breathing Document

No policy stays the same forever. It's a living, breathing document that evolves over time. Everything in life changes. In fact, change is the only certain thing in life. So as our urban circumstances change and new, useful information becomes available, our policies should evolve accordingly.

The best policies out there are those that are flexible enough to adapt to the times while staying true to their original intent. And there's always room for improvement. Remember that protopias are always spiraling upward, always growing, always improving—as should policy.

While policy might seem dull or overwhelming at first glance, it is actually one of the most powerful tools in regenerative placemaking. It transforms bold, visionary ideas into tangible, lasting change that benefits both people and the environment. However, policy shouldn't be static; it must evolve, adapting to the needs of the place.

Take Minneapolis, for example, where single-family home zoning was banned in urban areas to combat rising housing costs caused by low-density development. Density is our destiny in cities, but it must be approached thoughtfully through smart growth. By actively engaging with the policymaking process, we can shape the communities we want to live in, ensuring they remain accessible for future generations.

The more we understand and engage with the policymaking process, the more effective we can be in shaping the communities we want to live in.

Global Issue Areas for Regenerative Placemaking Policy

Effective policies can be applied universally, no matter where in the world you're placemaking. What ties these global issue areas together—and makes them effective for regenerative placemaking—is that they are all rooted in the same core philosophy: designing for community, nature, and culture.

Housing

The housing involved in a regenerative placemaking project should be affordable, sustainable, and inclusive. Everyone deserves a roof over their head that doesn't cost them the earth. Policies here should consider not only the physical structures but also how housing fits into the broader place. Housing is more than just shelter; it's about building upon the three pillars: community, nature, and culture. There's always a way to integrate all three.

Transportation

The only option for getting from point A to point B within a city shouldn't be inside a car that guzzles gas and has people stuck in traffic for hours. Think bikes, scooters, buses, subways, and trains—all offering a lot lower carbon emissions. Consider the 15-minute city concept, championed by Carlos Moreno and exemplified by cities like Paris, which aims to eliminate cars in the urban core. This approach ensures that all essential services and amenities are within a 15-minute walk or bike ride, creating a more connected, human-centric urban environment. As Moreno explains, the 15-minute city is about building human-scale proximity, where the most important activities in people's lives are accessible within a quarter hour by foot or by bike.[2]

But remember, a bus system, for example, should not be created in a vacuum; it's part of a wider system that enhances access to jobs, schools, hospitals, and cultural activities. It's about integrating transportation within the living whole of a community, ensuring it supports the local ecosystem and sustainable, connected living.

Energy

Solar panels on everything. Wind turbines on every hill. Biofuel tanks in every garden. This is overly simplistic, but you get the idea.

Transitioning to renewable energy is like upgrading from a gas to an electric car: It's inevitable; just give it time. Policies here should ensure energy systems are at the cutting edge of sustainable living and integrated into the broader community context, making energy more affordable in the long term and reducing carbon dioxide emissions and, therefore, global warming. This is where a regenerative placemaking community can do its bit to keep the planet from turning into a giant, uninhabitable toaster.

Waste Management

Use "Reduce, reuse, and recycle" as a collective mantra in your regenerative placemaking project. Waste management policies should include public trash cans, recycling stations, communal composting, and initiatives to minimize single-use plastics and other nonrecyclable materials. Consider implementing zero-waste policies and encouraging upcycling, which can transform waste into valuable resources. Recently, at New York Fashion Week, the first recycled clothing fashion show took place, showcasing how waste can be creatively repurposed. Waste management isn't just a boring, compulsory task; it can significantly enhance a community's quality of life, physical health, and culture, and the beauty of a place.

Water

This is for clean and safe consumption, first and foremost, but not just for drinking. Policies here make sure that water is used wisely, from the tap to the community gardens. Water policies should consider the entire ecosystem of a place, ensuring that they support local agriculture, public spaces, and the overall health of the community.

Public Spaces

Parks, plazas, community centers, and pedestrian pathways should be places where people want to spend time—not just somewhere they cut through on the way to somewhere else. Effective policies here involve expanding these spaces and considering who owns them, ensuring they remain accessible and beneficial to all. Participatory planning is key, allowing the community to help design spaces that reflect local culture and needs. This approach creates public spaces that are fit for purpose and serve as true communal areas, bringing about a vital sense of shared ownership.

Local Economy

Supporting local businesses is like watering your plants—they'll grow, thrive, and add color to any regenerative community but only with proper care and attention. In regenerative placemaking, the economy isn't just about money flowing in; it's about who it flows to, how it circulates, and what kind of legacy it leaves behind. That's why we believe local economic development should go hand in hand with dignity, creativity, and sovereignty.

Policies that support dignified jobs, small businesses, creative industries, and alternative education programs are a powerful foundation. But now, with the rise of blockchain and Web3 technologies, we have a new set of tools to deepen this mission—and potentially rewire the economy to favor local ownership, transparency, and inclusion from the start.

Imagine if the people who build the vibrancy of a neighborhood—artists, entrepreneurs, teachers, youth mentors, healers, farmers—could also own a piece of it. Future of Cities* is exploring how this is possible through blockchain-based technologies like tokenization, DAOs, and fractional ownership. These innovations allow a broader group of people—not just wealthy investors—to own small shares in developments, cultural spaces, or housing initiatives. We call this *profit participation*, and it means that those who contribute value can also receive value in return.

For example, in our PHX-JAX project in Jacksonville, we're exploring how blockchain can be a tool to create a decentralized community governance system where decisions about land, housing, and investment can be made transparently, with voting power distributed across local residents, not just developers or politicians. We're also exploring how tokenized real estate and digital assets can be used as a way to secure land titles, protect cultural memory, and help local creators earn revenue.

One of the more exciting applications is a community cryptocurrency, or *fungible token,* that allows people to transact locally, build credit within the community, and even access rewards for participation in local initiatives. This kind of local currency can be designed so that a portion of every transaction goes directly into a community housing or arts fund—fueling projects that benefit everyone.

Of course, there are risks with any emerging technology, and blockchain is no exception. That's why we're approaching this carefully, ensuring that our strategies are cocreated with community input, rooted in transparency, and aligned with environmental responsibility. We're using carbon-friendly protocols, running education programs to build local knowledge, and leaning into the *seventh-generation* principle—making decisions today that will still benefit the community seven generations from now.

Back in Wynwood, Miami, when large corporations like Walmart wanted to set up shop, the community pushed back and demanded terms that supported local businesses. It showed us what's possible when communities come together to protect their economic fabric. Today, with blockchain in our toolkit, we can go even further—embedding local empowerment and ownership into the infrastructure of how places are built, governed, and sustained.

This example highlights the importance of participation from all socio-economic and cultural backgrounds to be holistic in creating a truly regenerative community. Initiatives like these open doors, build bridges, and grow local economies, ultimately improving the quality of life for the entire community.

Food Systems

Farm to table isn't just a trendy phrase; it's about making a real commitment to sourcing food locally wherever possible. Eliminating large-distance supply chains, reducing waste, and focusing on local and seasonal goods is more sustainable and better for everyone, as well as for the environment. As a regenerative placemaker, you should ask yourself how a community can get access to food that's not just nutritious but also affordable, with a special focus on sourcing locally to benefit both local sellers and the planet. These policies should create win–win–wins for health, community, and the environment.

Health and Well-Being

This goes beyond just making sure your community doesn't get sick from polluted water and air. Rather, it's about creating environments that promote holistic mental, physical, and social well-being—because happiness should be policy too. Health policies should be integrated into all aspects of urban planning, from transportation to public spaces, ensuring that all components of the place contribute to overall well-being.

Education and Awareness

Under this umbrella, we have policies that ensure everyone knows the importance of community, nature, and culture and has access to the education around them. This could be seen as an investment in the next generation of compassionate community members, ecowarriors, and cultural protectors. This involves not just formal education but also alternative, creative, and fun learning opportunities that align with the needs of the local economy and community aspirations.

Resilience and Adaptation

These policies ensure that communities are prepared for environmental, social, and economic challenges. Because climate change, for example, is already here, policies need to ensure that communities are not just surviving but thriving in the face of it due to their collective resilience. This means integrating resilience into a lot of policy domains, from the storm- and flooding-proof structure of housing and public spaces to local food sources.

Although these are the primary issue areas, that's not to say that some policies won't overlap between categories. These are simply umbrella categories to make it easier to understand. Depending on what your regenerative placemaking project entails, it may well mean that some of these policy categories don't apply. For example, if you're working on a small community garden, you'll probably not need to focus heavily on transportation or housing policies. Instead, you would concentrate on policies related to land use, waste management, and food systems to ensure that your garden contributes positively to the local environment. On the other hand, if your project involves revitalizing a neighborhood with mixed-use developments, you'll likely need to address a broader range of policy areas, such as housing, transportation, public spaces, and the local economy.

Remember, regenerative placemaking isn't limited to large real estate projects led by developers; it can be initiated by anyone, anywhere. The key is to assess your project's unique needs and context, picking and choosing which policy areas are most relevant and where overlap might occur. This holistic approach will help you create a strategy that aligns with the goals of your regenerative placemaking project. That said, remember to choose a wide range of policies that support at least one of the pillars of community, nature, and culture.

CONCLUSION

Emerald Cities

There's no place like home.[1]

—DOROTHY GALE

Dorothy had come such a long way. She was exhausted but triumphant. She'd learned everything she needed to learn to, finally, at long last, return home to a place she belonged.

She closed her eyes, tapped her ruby-red slippers together, and whispered those well-known words, "There's no place like home." Then, she suddenly got an insight that hit her like a ton of bricks, one that would change her perception forever: The home she was looking for was never outside herself. It was never in a far-off place and never solely to be found in the superior hands of a powerful wizard like she had presumed it was. Everything she could have ever wished for was within *her* all along, just as the Scarecrow had always had his wisdom and intelligence, the Tin Man his heart, and the

Cowardly Lion his courage. She had the power to envision a home she knew she belonged in, and she had the power to create it.

They all realized that their journey together wasn't just about reaching the Emerald City (although it was indeed stunning); it was about coming together to work toward a shared goal, a collective dream that would bring their deepest wishes to reality.

You see where I'm going with this. Maybe before you read this book, you would have been skeptical about the comparison between regenerative placemaking and a classic yet totally trippy movie brought out in 1939. But the same lessons learned along the Yellow Brick Road are the same lessons we're learning now as modern-day regenerative placemakers. No matter how long the path may be, we have everything we'll ever need to arrive at our protopia if we learn to work together. (If you're a millennial or Gen Z and have absolutely no idea what I'm talking about, consider watching this movie as your pop culture homework. I promise it's worth the weirdness—flying monkeys and all.)

As regenerative placemakers, we're aiming for Emerald Cities in whatever place we're working on. The Emerald City isn't just a fantastical fictional fantasy; it's a symbol of regenerative placemaking in action. Our vision for the future is green, balanced, and beautiful. We envision a place where we mimic nature in every part of urban life—metropolises with green streets, green parks, green roofs, and green walls—and where built environments are autonomous, adapting to their surroundings and responding to environmental changes. We don't want to live in cities without sunlight or green spaces, and we don't want to live without a green, regenerative-minded community by our side either. In our ideal future, cities thrive in harmony with people and nature, blending technology and ecology to create environments that are regenerative. That's our vision of the Emerald City. It's not the one you saw in the movie; it's even better.

It's Not About Slowing Down Entropy

Remember this: A regenerative approach isn't just about slowing down entropy, preventing disorder, and attempting to avoid degeneration. Our mission goes beyond that, and it also goes beyond sustainability, which, in reality, is only one step up from entropy. Regenerative placemaking shifts the focus from merely sustaining the status quo to enhancing the ability of living communities and ecosystems to thrive and evolve. That's what regeneration stands for, and that's what all of us must aim for.

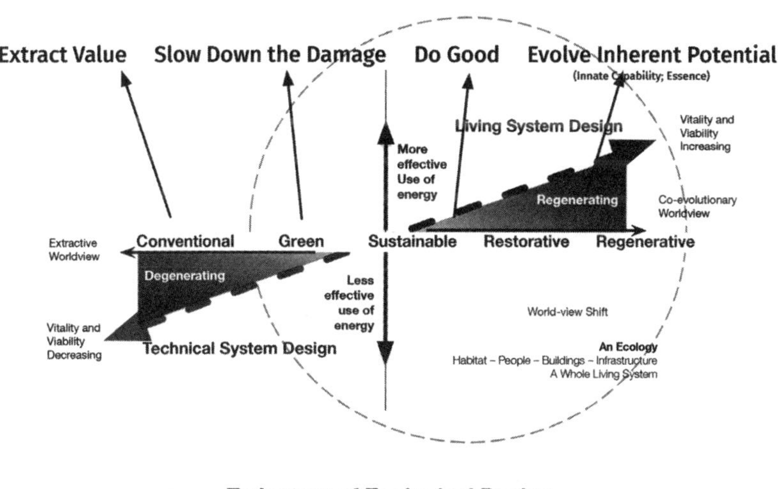

Trajectory of Ecological Design by Bill Reed & Regensis Institute, from degeneration to regeneration 2000–2024

As the Regenesis Group says, it's not enough to just maintain what already is. We're therefore invited, as placemakers, to facilitate and "build the capability of living communities to evolve toward greater value."[2]

This approach isn't optional. It's essential. Even broader global frameworks like the seventeen Sustainable Development Goals (SDGs) created by the UN and UNICEF are beginning to mirror this mindset.[3] The UN SDGs are a good example of frameworks that used to focus solely on statistics and tangible goals. Now, these frameworks are evolving to reflect a deeper understanding of what true, holistic progress entails. As a result, the UN is rethinking its approach and is now launching a brand-new framework that is more aligned with regenerative work, recognizing that the health of our planet and the health of its people are deeply interconnected—a profound yet forgotten truth that we are choosing to remember. The representatives of the UN are actually regenerative placemakers too, whether they know it or not, who acknowledge the importance of restoring and rejuvenating natural systems, as well as building resilience, well-being, dignity, and a sense of togetherness within communities.

To fully embrace regenerative placemaking, we must go through a significant shift in mindset—in how we think and how we act. Let us move away from competition toward collaboration, and from consumption toward meaningful contributions.

Regenerative placemaking is about transformation from the inside out. In this work, we refer to the process of improving ourselves as *developing individual capacity*. The idea is that, by enhancing our own skills and abilities, we contribute to a collective effort to unlock the shared potential of both people and places.

By investing in each person's unique strengths and empowering them to grow, we elevate the entire community. This growth isn't just personal; it amplifies the collective power, allowing us to tap into the creative energy necessary to shape our environments into vibrant, thriving communities. It's a holistic approach where both people and places flourish together, interwoven and interdependent, driving meaningful change for the future. When we work together, as a conscious, collective force for good, *anything* is possible. And it starts with embracing a regenerative approach to everything we do as placemakers.

As Dr. Dominique Hes emphasizes, this shift involves seeing regenerative development and design not as a tool for solving problems but as a means of cocreating potential and synthesis that brings together all parts of a system to create a greater whole and a conscious sense of oneness.[4] The bottom line is this: We're at the forefront of the regenerative mission, we're in this together, and we'll never be alone in our mission.

What Now?

Now that you understand the theory and experiential framework behind regenerative placemaking and how it serves the greater whole, you're ready to step out into the world and start making a real difference. As an individual, you have the power to create some positive change, but as part of a team, your impact can be exponentially bigger.

Throughout my own journey in regenerative placemaking at Future of Cities®, partnerships have been vital. Organizations like the

Regenesis Institute and PlacemakingX have been invaluable allies, offering support and shared wisdom as my team and I navigated the complexities of urban development and community building. These collaborations have allowed us to refine our approaches, creating more impactful and sustainable projects.

To support you in your journey, I encourage you to tap into the wealth of resources and training Future of Cities® and these other amazing organizations have to offer. The Regenesis Institute's Regenerative Practitioners Series provides a really deep training I completed myself in 2024 and offers an official qualification in regenerative development, if that's something you believe would serve you. PlacemakingX also offers a variety of training and courses that delve into effective placemaking strategies and techniques.

At Future of Cities®, we're always up for collaboration with those who share the vision of a regenerative future built on the pillars of community, nature, and culture. We're coleading a global community costewarding the regenerative development movement, and of course, we'd love you to be part of it. We also run events all around the world that bring together passionate individuals and groups who work hard and play hard in the name of regenerative placemaking, as well as fun city-making festivals and summits. Go to our website at focities.com to find out more.

So, connect, collaborate, and continue to inspire change. Don't forget that the journey of regenerative placemaking is one of constant evolution, growing in innovation and clarity at an unprecedented rate. This means it's more important than ever to stay informed and to build strong partnerships so we can navigate these waters together.

We Are the Pillars

If you take anything away from this book, let it be this: The essence of regenerative placemaking lies in recognizing that the true power to hold up the pillars of community, nature, and culture is within all of us. It is within our shared purpose and commitment to nurturing the places we call home. The pillars you've read about aren't ideas that are leaned upon for inspiration.

We *are* the pillars. You, reader, are bringing life to them right now, just by reading this book. Each one of us can contribute to the strength and stability of our communities, the resilience of our natural environments, and the vibrancy of our cultures. Regenerative placemaking never has, nor ever will be, a solitary mission; it is a collective effort that requires everyone to lean in, listen, and act with intention. The pillars of community, nature, and culture are far from mere theoretical concepts; they are alive in the choices we make every day.

When we invest in our people, we strengthen the pillar of community. When we protect our natural world, we strengthen the pillar of nature. And when we honor and celebrate our diverse cultures, we strengthen the pillar of culture. We mean it when we say that this isn't a vision for the future but a call to action for the now. And it's a call to action that allows the extraordinary potential that lies within us all to come to the surface.

The truth is that there's a Dorothy in all of us, most of all in regenerative placemakers. And you're one of us now. No matter who you are or where you come from, no matter how big or small your role to play is, trust me when I say that without you, arriving at the Emerald City of all our dreams will be a lot harder. Whether you are a community organizer, an artist, a city planner, a gardener, or simply a neighbor who cares, your voice and actions matter more than you know. Your love for the place you call home, your willingness to engage, and your commitment to positive change are what make regenerative placemaking possible.

This vision we all share at heart, the one where all life on Earth lives harmoniously, joyfully, regeneratively, and compassionately, is going to take everybody's best effort. Genuinely ask yourself these questions: *What can I do* today *to make a positive impact on the place I call home?* That's to say, *How can I support my community? What steps can I take to protect the animals and plants that share our neighborhood, to care for the water we drink and the air we breathe? How can I bring people together to celebrate the homegrown culture that makes our community unique?*

It's going to take the leader you have deep within to come up and out of hiding in the name of the collective good. It's going to take courage and strength. But more than all of that, it's going to take a huge heart. Because at the heart of regenerative placemaking is love—love for the communities we serve, love for the natural environments we strive to protect, and love for the diverse cultures that paint our world with color.

I hope that, by now, this is clear: Regenerative placemaking isn't just sustainable urban planning. It's about the connections between each and every one of us. We all deserve to live in environments that are safe, uplifting, and reflective of who we are—places that support us as we evolve and take our world from degradation to regeneration.

I want to thank you for reading this book and opening yourself up to a lifelong pilgrimage to making the world a better place. I'm excited to walk this path with you to the Emerald City. The journey may be long, and there will be obstacles along the way, but with every heart united in purpose, a regenerative world isn't a fantasy; it's a place we can all reach and live happily ever after in together. Thank you for being by my side as we all strive for a collective protopia on Earth. Let the journey begin.

TOOLKITS

TOOLKIT 1

Community

By now, you've grasped the crucial role that community plays in regenerative placemaking. Deep listening is a crucial tool in this process. It means listening properly, on and between the lines, to truly understand the messages the community is sending you. But if deep listening is the key, we need to know what questions to ask to get those meaningful answers we can take action on.

Communities often don't have a crystal-clear picture of what they need or how to turn their visions into reality right off the bat. If you're new to a varied team of regenerative placemakers, figuring out how to draw out insights from the community can be tricky at first. That's where this toolkit comes in handy.

Once you've set up a welcoming environment, scheduled various times for community members to attend, and made it easy for everyone to join, it's time to greet your community warmly. You can introduce your meeting as a *kitchen conversation*, one that's open and informal but that will be taken seriously. Introduce yourself in a humble way so nobody picks up on unwanted power dynamics. Clarify your intentions, and emphasize that your goal is to serve their needs. Ensure the community that their voices matter now more than ever and that your role will be to ask the right questions and listen.

With that foundation laid, you can use this toolkit to guide your session. These questions are designed to elicit detailed and honest contributions. Consider the following questions as part of your upcoming meaningful conversations with the community:

What are the current challenges you face in your neighborhood?

- **Why it matters:** Understanding the community's pain points can help prioritize which issues to address first. Make sure to show some empathy here and don't interrupt.
- **Follow-up question:** "What challenges has your neighborhood faced in the past?" Make sure you stay tuned to gather any information on what developmental mistakes have been made previously so you can avoid them this time around.
- **Example community answer:** "We have issues with poor street lighting and littering in public spaces. There's also a lack of affordable childcare options for working parents."

What aspects of your neighborhood do you value the most? Where do you take loved ones or friends when they come to visit your neighborhood?

- **Why it matters:** Identifying cherished elements of the place can guide you on how to protect, preserve, and enhance these features. Make sure to validate and celebrate these aspects.
- **Example community answer:** "I love the beautiful community garden that brings people together."

What kind of changes would you like to see in your community?

- **Why it matters:** This helps uncover residents' aspirations and visions for their ideal neighborhood. Let ideas flow freely, take notes, then paraphrase them. Ask them if you have understood their desires properly.
- **Example community answer:** "I'd like to see more facilities for kids and improved public transportation options. Also, some of the old, vacant buildings could be renovated into useful community spaces."

Are there specific local businesses or services that you feel are lacking?

- **Why it matters:** This can reveal gaps in amenities and opportunities for local economic development. Explain that their ideas won't just improve their quality of life, but provide a lot of job opportunities to community members. Tell them to not hold back, then, once you've listened, identify the common thread and focus on what the majority claim they're lacking.

- **Example community answer:** "There's no good grocery store nearby. We could really use a place that offers fresh produce and other essentials. Also, a café or bakery would be great for social gatherings."

How do you see the role of public spaces in your community?

- **Why it matters:** Understanding preferences for parks, community centers, and other public areas can shape the design of these spaces.
- **Example community answer:** "Public spaces should be safe, clean, and accessible to everyone. They could serve as venues for community events, farmers' markets, and casual meetups. A well-maintained park with playgrounds and sports facilities would be ideal."

What types of community events or activities would you like to see more of?

- **Why it matters:** Insights into preferred events can give you a window into the collective culture of the area that already exists and into what is on the horizon. Make sure they understand that events and activities can become the glue of a community and the best way to enhance engagement and avoid loneliness.

- **Example community answer:** "It would be fantastic to have more cultural festivals and outdoor movie nights. Regular farmers' markets and craft fairs would also be great and allow us to sell our produce."

How can we ensure that new developments or changes benefit everyone in the community?

- **Why it matters:** This question seeks to address concerns about equity and inclusivity in planning processes.
- **Example community answer:** "Let's not forget about the children that live here. We should make sure that there are spots for kids to play, as well as affordable housing options, and facilities that can be easily used by those with physical disabilities."

What kind of resources would help you thrive in your neighborhood?

- **Why it matters:** Identifying needs for education, employment, or other resources can help tailor development projects to actual community requirements. Here's a basic list for you to provide examples if the answers don't roll off their tongues:
    - **Education and training:** After-school programs and job training?

- **Employment support:** Job fairs and career center services?
- **Health and wellness:** Affordable health services and mental health support?
- **Affordable childcare:** Reliable and accessible childcare options?
- **Cultural and recreational activities:** Community events, art workshops, and sports leagues?
- **Access to technology:** Technology and internet access?
- **Local food resources:** Community gardens and farmers' markets?

- **Example community answer:** Listen out for the above and categorize the responses, as well as making room for completely new suggestions.

What are your concerns about potential changes or new developments?

- **Why it matters:** Understanding fears or objections helps in addressing potential resistance and finding solutions. Then, reassure the community that you'll do everything in your power to make sure these fears don't become a reality.
- **Example community answer:** "I'm worried that new developments might make the area unaffordable for longtime residents. I also fear that the spirit of our neighborhood could be lost."

How can we best communicate with you and keep you informed about upcoming projects?

- **Why it matters:** Ensuring effective communication channels creates and builds trust and keeps the community engaged in the development process. Assure them that this meeting wasn't just a one-time thing, and that they will be needed, valued, and listened to throughout the entire process.
- **Example community answer:** "Regular updates through community newsletters and social media would be great. Hosting town hall meetings and having a dedicated website for project information would also help keep everyone informed and involved."

Afterward, take a moment to express your sincere gratitude for their invaluable input. Acknowledge the time and effort they've invested in sharing their thoughts and ideas. Emphasize how crucial their feedback is for the project's success and encourage them to continue contributing their insights throughout the development process.

Once the right kind of information is collected, it's vital to ensure it is effectively used. Clearly communicate how their feedback will directly inform the decision-making process of the policy or project. Explain how it will shape the policy, its objectives, and methods, and detail how their insights will influence the design, functions, and implementation strategies of the project. By doing so, you reinforce the value of their contributions and demonstrate that their input is not just heard but actively shapes the project's direction.

Also, make sure that they have multiple avenues to provide ongoing feedback, whether it's through online platforms, written surveys, or face-to-face meetings. By offering various methods of communication, you make it easier for them to stay involved and voice their opinions, as well as showing them you genuinely care about getting everyone involved. Remind them that the success of your project hinges on their active engagement and participation.

A really nice addition to the service you provide would be a visual of the main takeaways from the meeting or meetings that you hold. This visual could be shared with every community member who attended, as well as those who couldn't make it, through various channels such as email, postal mail, or posters in public spaces.

Following is an example of the visual we created from PHX-JAX community outreach meetings, which was shared both internally with the project team and externally with the residents and participants. We also chose to provide contact information so that the community members could reach out to us with any questions or further contributions. Ultimately, remember that success moves at the speed of trust and that nurturing a strong, trusting relationship with the community is key to achieving success in your regenerative placemaking project. Often, it's the small gestures that make the biggest difference.

TOOLKIT 2

Nature

When embarking on a regenerative placemaking project, understanding the existing natural environment is very important. This toolkit provides key questions to help you assess and engage with the local ecosystems effectively. Use these questions to guide your investigation and identify who to consult for deeper insights. It goes without saying that you can always ask the community members in your kitchen conversations, but the answers you receive will probably be incomplete. For more detailed answers and statistics, you can consult the "Who to Ask" sections that follow.

What are the native species of plants and wildlife here?

- **Why it matters:** Identifying native species helps you understand the local biodiversity and its role in the ecosystem.
- **Who to ask:** Local environmental organizations, wildlife experts, or native plant nurseries.

What can this biodiversity offer to our project?

- **Why it matters:** Identifying native species is essential to understanding local biodiversity and its role in the ecosystem. Native plants are naturally adapted to the area and often require fewer resources—like water and fertilizer—compared to nonnative species or traditional grass lawns, which are resource-intensive. Once established, native plants and edible species are typically more resilient, thriving with minimal intervention. By grouping plants into guilds—communities of mutually beneficial plants that support each other's growth—you can further enhance resilience. Plant guilds work together to improve soil health, retain moisture, deter pests, and create a self-sustaining ecosystem that requires minimal upkeep.[1]
- **Who to ask:** Ecologists, landscape architects, or conservationists.

How could my project potentially hurt or damage the environment?

- **Why it matters:** Understanding potential negative impacts helps you mitigate harm and plan for conservation.
- **Who to ask:** Environmental impact assessors or local conservation groups.

What are the benefits of engaging with water, land, and air in clean, sustainable, and renewable ways?

- **Why it matters:** Sustainable practices protect ecosystems and enhance project longevity.
- **Who to ask:** Sustainability consultants, environmental engineers, or local government environmental departments.

How have water, land, and air been managed so far?

- **Why it matters:** Knowing past management practices helps identify current conditions and areas for improvement.
- **Who to ask:** Local environmental agencies, historical records, or community elders.

What has been done incorrectly (e.g., damaging or polluting), and what has been done well?

- **Why it matters:** Learning from past mistakes and successes guides better decision-making and management.
- **Who to ask:** Environmental historians, local community members, or environmental advocacy groups.

What are the relationships between the natural ecosystems and human systems of the place?

- **Why it matters:** Understanding these interactions helps integrate natural and human systems effectively.
- **Who to ask:** Urban planners, social scientists, or local community organizations.

How can these relationships be improved?

- **Why it matters:** Enhancing interactions between natural and human systems supports sustainability and resilience.
- **Who to ask:** Environmental consultants, community engagement specialists, or local government officials.

What nature-based solutions are available here?

- **Why it matters:** Nature-based solutions can address environmental challenges while benefiting the ecosystem.
- **Who to ask:** Ecological restoration experts, nature-based solution specialists, or local environmental organizations. Understanding the cultural landscape is essential in regenerative placemaking.

What is the potential for this place to be healthy, and what will that look like?

- **Why it matters:** Establishing a shared vision of health for the area serves as a North Star, creating a unifying aim that brings together diverse interest groups. When each group applies its unique knowledge toward this common goal, it breaks down problem-solving silos and encourages integration.
- **Who to ask:** Local health advocates, environmental consultants, community members, and interest groups with expertise in ecology, urban planning, and public health.

This toolkit provides key questions to help you explore and engage with the local and broader cultures. Use these questions to guide your research and identify who to consult for deeper insights. Using this toolkit helps you understand the natural environment and how it relates to your project. This way, you can make sure your work is both relevant and effective. Remember, regenerative place-making is not a one-size-fits-all approach, ever. So be prepared to be patient, get equipped with the right questions, and deeply listen!

TOOLKIT 3

Culture

Understanding the existing culture of a place is paramount. This toolkit provides key questions to help you assess and engage with the local culture effectively. Use these questions to guide your investigation and identify who to consult for deeper insights. After all, as the famous saying goes, "culture eats strategy for breakfast,"[1] so understanding it is crucial to achieving any lasting success.

What are the local, regional, national, and international identities and cultures here?

- **Why it matters:** Identifying the diverse cultures present helps ensure that your project respects and reflects the whole community.
- **Who to ask:** Local cultural organizations, historians, or community leaders.

What are the arts, music, cuisine, entertainment, sports, literature, dance, theater, dining, etc. present in this place?

- **Why it matters:** Understanding the cultural expressions in the area helps you incorporate them into your project, making it more relevant and engaging.
- **Who to ask:** Local artists, cultural centers, or community members.

What languages are spoken?

- **Why it matters:** Language diversity is a key cultural element that can influence communication and engagement strategies.
- **Who to ask:** Local schools, linguistic experts, or community organizations.

What are the histories of migrations of populations?

- **Why it matters:** Knowing migration histories helps you understand the community's demographics and cultural influences.
- **Who to ask:** Historians, local archives, or elder residents.

What diasporic communities exist here?

- **Why it matters:** Identifying diasporic communities (groups of people who have spread or been dispersed from their original homeland and who maintain connections to their cultural, linguistic, or national heritage) can reveal unique cultural assets and opportunities for engagement.

- **Who to ask:** Community organizations, cultural associations, or social researchers.

What opportunities exist to amplify cultural assets?

- **Why it matters:** Amplifying cultural assets can strengthen community identity and support the success of your project.
- **Who to ask:** Cultural planners, local artists, or community leaders.

How can my project engage in listening deeply to Indigenous wisdom, knowledge, and practices?

- **Why it matters:** When appropriate and relevant, engaging with Indigenous wisdom ensures that your project respects and integrates traditional knowledge and practices.
- **Who to ask:** Indigenous leaders, cultural advisers, or elders.

What landmarks exist, and what is the story of each heritage site?

- **Why it matters:** Understanding the significance of local landmarks and heritage sites ensures they are honored and preserved in your project.

- **Who to ask:** Local historians, cultural heritage organizations, or tour guides.

What are important dates in recent events (e.g., natural disasters, new universities, political events)?

- **Why it matters:** Awareness of recent events helps you understand the community's current context and sensitivities.
- **Who to ask:** Local news archives, community leaders, or government records.

By using this toolkit, you can better understand the cultural dynamics of the area, allowing your project to resonate more deeply with the community. Get creative and ask as many questions as you can. And as always, make sure you're listening more than you're talking!

TOOLKIT 4

Metrics

In regenerative placemaking, balancing fixed metrics with community-driven ones is key. Fixed metrics, which you'll find in the following, provide essential life signs for stakeholders, delivering data on the impact, sustainability, and financial health of the project. These metrics help ensure accountability and demonstrate progress on a larger scale.

What truly sets regenerative placemaking apart, however, is allowing the community to set and track their *own* metrics. When community members define what success looks like for them—whether it's an increase in local art spaces, improved social cohesion, access to green areas, or health resources—they feel a greater sense of agency and ownership over the project. These community-driven metrics reflect what residents find most meaningful, encouraging active participation and investment in the process.

This approach also facilitates long-term care and connection to the place, as people are empowered to measure and witness positive changes in areas they value. Similar to the Indigenous concept of totems, which serve as symbols of cultural values and relationships to the land, community-set metrics root the project in the needs and identity of those who live there. This dual metric structure—fixed indicators for accountability and community metrics for empowerment—creates a project that is both grounded and adaptable, with true regenerative potential.

With that in mind, feel free to use metrics that apply at different levels of scale. Some metrics may be best suited for the project level, while others are more impactful at the city level, addressing broader, spatially layered systems. By approaching metrics this way, we not only capture the project's immediate impact but also integrate it within larger citywide and regional dynamics, maximizing the reach and effectiveness of regenerative goals.

People (Community and Culture): Social Metrics

Participation in the Project

- The percentage of community members participating in public consultations or community meetings
- The percentage of meetings held per location accessible to the relevant neighborhoods
- The percentage of community members who feel their views are considered in project decision-making
- The percentage of community-led initiatives supported by the project

Mental Health

- Community mental health outcomes (e.g., rates of depression, anxiety, or other mental health issues and anecdotal evidence of increased well-being)

Physical Health

- Access to healthcare services in the area
- Physical health outcomes (e.g., rates of chronic diseases decreasing, life expectancy longer)
- Availability of recreational facilities (e.g., parks, gyms, sports complexes)
- Participation in physical health programs (e.g., community fitness classes, wellness programs)

Social Cohesion

- Participation of community members in community activities and events
- Number of shared community spaces
- Social capital measures (e.g., trust in local institutions, neighborhood safety, crime rates)

Accessibility

- Infrastructure accessibility (e.g., compliance with the Americans with Disabilities Act, public transportation availability)

- Access to essential services (e.g., grocery stores, banks, post offices)

Education

- Access to educational resources (e.g., the number of schools, literacy rates)
- Participation in lifelong learning programs
- The number of workshops and educational events held

Civic Engagement

- Voter turnout rates in local elections
- Volunteerism and community service participation rates

Diversity and Inclusion

- The representation of minority groups in community gatherings, events, and leadership positions
- The inclusion in decision-making processes (e.g., the percentage of community members involved in planning)

Local Sourcing

- Local sourcing ratio (the percentage of goods and services sourced from local businesses)

Affordable Housing

- The percentage of housing units designated as affordable
- Housing stability metrics (e.g., eviction rates)

Cultural Participation

- The number of cultural events
- Attendance rates at cultural events
- Membership or participation in local cultural organizations

Preservation of Heritage

- The number of historic sites or cultural landmarks preserved
- Investment in cultural heritage projects

Artist Residency

- The number of artist residencies or cultural exchange programs

Cultural Funding

- Total dollars allocated to arts and cultural grants

Creative Economy

- Economic contribution of the creative sector (e.g., job creation, revenue generation)
- The growth rate of creative industries (e.g., film, music, publishing)

Public Art Installations

- The number of public art projects or installations
- Community engagement in public art initiatives

Local Media

- The presence and reach of local media outlets that promote cultural content
- Representation of diverse cultural voices in media

Cultural Tourism

- The impact of cultural tourism on the local economy
- Visitor satisfaction with cultural experiences

Planet: Environmental Metrics

Climate Resilience

- The implementation of climate adaptation strategies (e.g., flood defenses, strong infrastructure, drought-resistant crops)
- Community resilience scores (e.g., ability to recover from climate-related events)

Carbon Footprint or Emissions to Air or Water

- Emission levels of key air pollutants (e.g., NO_x, SO_x, PM2.5) and water quality (e.g., chemical oxygen demand, presence of contaminants)
- The percentage of renewable energy used like wind, solar, or hydroelectric power

Green Spaces

- The number of green areas preserved, introduced, or reforested—using the Living Building Challenge or other green building rating systems would support this
- The impact on local biodiversity (e.g., species diversity index)

Permaculture and Food Self-Sufficiency

- The percentage of the community or project area engaged in growing their own food
- The number of food forests, gardens, urban farms, or allotments
- The amount of food produced locally (e.g., pounds of produce harvested annually)

Waste Management

- The waste diversion rate (percentage of waste recycled or composted)
- A reduction in single-use plastics or nonrecyclable materials

Energy Efficiency

- Energy-use intensity metrics
- The implementation of energy-saving technologies (e.g., LED lighting, smart meters)

Sustainable Transportation

- The percentage of trips made by car, public transport, cycling, or walking
- The number of electric vehicle charging stations

Profit: Economic Metrics

Local Job Creation

- The number of jobs created directly and indirectly by the regenerative placemaking project
- The percentage of jobs filled by local residents

Economic Resilience

- The number of new local businesses established or scaling of current businesses
- The business survival rate over time

Return on Investment

- The financial return relative to project costs
- The social return on investment considering broader community benefits

Tax Revenue

- An increase in local tax base due to the project
- Tax contributions from new businesses or increased property values

Economic Diversity

- The diversity of local industries (e.g., manufacturing, services, creative sectors)
- Dependence on a single industry or employer
- The number of entrepreneurs

Financial Inclusion

- Support for local entrepreneurship and small business development

Income

- Median household income growth

Affordable Commercial Space

- The availability and affordability of commercial real estate for small businesses
- Vacancy rates in commercial properties
- The number of coworking spaces and office spaces

Infrastructure Investment

- Dollars invested in local infrastructure (e.g., roads, utilities, public spaces)
- The impact on infrastructure quality and accessibility

Cultural Investment

- Money invested in local cultural zones and events

Tourism Revenue

- The economic impact of tourism (e.g., spending by visitors, hotel occupancy rates)
- Growth in tourism-related businesses

The list of possible metrics is extensive, and it's entirely up to you to decide how many to focus on. You have the flexibility to tailor your approach based on what's most important to your project, ensuring you track what truly matters for your success. Also, the specific metrics you'll use for your project might not be present in this list. You may find that some additional metrics are needed or that certain metrics listed here don't apply to the current stage of your project. That's perfectly fine. What's important is that you have a clear understanding of what aspects of PPP you'll be measuring from the outset. By defining these key metrics early on, you can stay laser-focused on designing and implementing strategies that align with your objectives, all of which contribute to the overarching goal: creating regenerative cities and neighborhoods of the future.

One crucial lesson in this process is that, in the world of placemaking, it's not enough to simply replicate what others have done. Instead, the focus should be on your unique approach and desired outcomes. Don't focus on the *what* of other successful projects but on the *how* of your unique approach. As I often tell those who ask if Little Haiti and PHX-JAX will become the next Wynwood, in order for a place to truly resonate, it needs to be an original—something connected to the people, the place, and the natural environment. Places are born original, and we want to keep them that way. It's great to be inspired by successful places like Wynwood, but it's even better to be unique, to let a place reflect its distinct culture and community.

Architect William H. Whyte, a hero in the placemaking movement, emphasized this point.[1] He discovered that when people tried to copy successful projects by mimicking their form—like design or layout—the results were often disappointing. The real key to success lies in the process: *how* a project is managed, *how* governance is structured, and *how* the community is involved and allowed to take ownership of the space. For both governments and the private sector, this lesson is crucial. The true success of placemaking isn't just about what you build but about *how* you manage the process.

TOOLKIT 5

Policy

This policy toolkit is organized according to the key policy areas outlined earlier. For each area, we provide three practical examples of policies that can be implemented to support regenerative placemaking efforts.

Housing

- **Affordable housing development:** Implement inclusionary zoning policies that require a percentage of new developments to be affordable for low- and moderate-income residents. This ensures that all community members have access to housing that is within their financial reach.
- **Sustainable building standards:** Adopt green building codes with incentives that promote energy efficiency, the use of sustainable materials, and the incorporation of renewable

energy sources in new housing projects. This policy supports long-term environmental sustainability and reduces operational costs for residents.

- **Mixed-use zoning:** Encourage mixed-use developments that combine residential, commercial, and recreational spaces. This policy facilitates walkable and cyclable neighborhoods that reduce dependence on cars and bring about community interaction.

Transportation

- **Complete Streets policy:** Implement a Complete Streets policy to design roads that are safe and accessible for all users, including pedestrians, cyclists, public transit riders, and motorists, along with some areas solely for pedestrian use. (For more information, revisit the "Policy Isn't Just for Governments" section of Chapter 7 or go to www.smartgrowthamerica.org/what-are-complete-streets.)
- **Public transit investment:** Increase funding for public transit infrastructure, including buses, subways, and bike-sharing programs, to provide efficient and affordable alternatives to car travel. This reduces traffic congestion and lowers carbon emissions. Consider car-free zones too.
- **Safe routes to school:** Develop policies that create safe walking and biking routes for children to get to school.

Energy

- **Renewable energy incentives:** Provide tax credits or rebates for residents and businesses that install renewable energy systems like solar panels and wind turbines. This encourages the transition to clean energy sources and reduces greenhouse gas emissions.
- **Energy efficiency standards:** Implement strict energy efficiency standards for new buildings and retrofits, ensuring that all structures minimize energy use and reduce their carbon footprint.
- **Community energy projects:** Support community-owned renewable energy projects, such as solar farms or wind cooperatives, that allow residents to generate their own power and benefit from collective energy savings.

Waste Management

- **Zero-waste policy:** Establish a zero-waste policy that encourages reducing waste at the source, reusing materials, and recycling. This includes banning single-use plastics and promoting other sustainable waste management practices.
- **Community composting programs:** Create local composting programs that provide residents with resources and facilities to compost organic waste. This reduces landfill use and creates nutrient-rich soil for community gardens.
- **Waste education initiatives:** Launch educational campaigns and workshops that teach residents about proper waste

sorting, recycling, and composting practices. This also builds a culture of sustainability and environmental responsibility.

Water

- **Water conservation ordinances:** Implement policies that encourage water-saving practices, such as rainwater harvesting, greywater reuse, and drought-tolerant landscaping. This ensures efficient water use and protects local water resources.
- **Stormwater management:** Use green infrastructure solutions, like permeable pavements and bioswales, to manage stormwater runoff and reduce flooding. This policy helps maintain healthy water ecosystems and protects communities from storm damage.
- **Rights of water:** Implement the rights of rivers, streams, oceans, and bays to be clean and protected. Ensure these vital waterways are free from pollution, safeguarding their health for future generations.

Public Spaces

- **Participatory design:** Create policies that require community involvement in the design and development of public spaces. This ensures that parks, plazas, and community centers reflect local needs and cultural values.
- **Green space requirements:** Mandate a minimum amount of green space in urban development projects to provide

recreational areas, improve air quality, and support biodiversity. No space? Get creative with green roofs and green walls.

- **Public art initiatives:** Fund public art projects and installations that enhance the aesthetic appeal of public spaces and celebrate local culture and heritage.

Local Economy

- **Small business support:** Provide grants, low-interest loans, and tax incentives for local entrepreneurs and small businesses. This policy stimulates local economic growth and creates jobs within the community.
- **Buy-local campaigns:** Promote buy-local campaigns to encourage residents to support local businesses and keep money circulating within the community. This strengthens the local economy and builds community resilience.
- **Community ownership:** Support the development of local ownership models using blockchain technologies, such as DAOs, community tokens, and fractional real estate investment. These tools allow residents, artists, and small business owners to own a financial stake in their neighborhood's future—creating more equitable, participatory, and resilient local economies.

Food Systems

- **Urban agriculture policies:** Support urban farming initiatives by providing land access, resources, and technical assistance to residents and organizations. This promotes local food production and improves food security.

- **Farmers' markets support:** Develop policies that encourage the establishment and growth of farmers' markets, providing local farmers with direct access to consumers and residents with fresh, healthy food options.
- **Food waste reduction:** Implement policies that reduce food waste by encouraging food donation programs, composting, and innovative solutions like food-sharing networks.

Health and Well-Being

- **Active living design:** Cocreate an urban design that encourages physical activity, such as walkable neighborhoods, bike lanes, and accessible public spaces. This promotes healthier lifestyles.
- **Mental health support:** Develop policies that provide accessible mental health services and create supportive environments for all community members, including public spaces designed to reduce stress and promote relaxation, like pocket forests.
- **Pollution control:** Implement strict air and water quality standards to reduce pollution and protect public health. This policy ensures a healthier environment for all residents (including animals and plants!).

Education and Awareness

- **Environmental education programs:** Integrate environmental education into school curricula and community programs to raise awareness about sustainability and ecological stewardship.
- **Community workshops and events:** Host workshops, seminars, and events that educate residents on topics such as energy efficiency, waste reduction, permaculture, and sustainable living practices. This helps to nurture a well-informed and engaged community.
- **Youth engagement initiatives:** Develop programs that involve young people in community planning and decision-making processes, to create a long-term sense of ownership and responsibility for the future of their community.

Resilience and Adaptation

- **Climate resilience planning:** Create and implement climate action plans that address vulnerabilities to climate change, including extreme weather events, sea-level rise, and heatwaves. This ensures communities are prepared and adaptable to future challenges.
- **Disaster preparedness training:** Establish policies that provide residents with training and resources for disaster preparedness and response, strengthening community resilience in the face of natural disasters.

- **Resilient infrastructure:** Invest in infrastructure improvements that enhance the resilience of buildings, roads, and public spaces to withstand environmental, social, and economic shocks.

This toolkit provides a range of policies across different issue areas to guide your regenerative placemaking project, but you must tailor these policies to your community's unique needs. Remember to involve stakeholders in the decision-making process to ensure that the solutions are inclusive, sustainable, and effective.

Notes

Introduction

1. 1000 Friends of Florida, Center for Landscape Conservation Planning, *Florida's Rising Seas: Mapping Our Future* (1000 Friends of Florida, 2024), https://1000fof.org/sealevel2040/wp-content/uploads/2023/06/Sea-Level-2070-Report-FINAL.pdf.
2. NASA, "Evidence," *Explore* (October 23, 2024) https://science.nasa.gov/climate-change/evidence.
3. Joe McCarthy, "There Could Be 2 Billion Climate Change Refugees by 2100," *Global Citizen* (June 29, 2017), https://www.globalcitizen.org/en/content/2-billion-climate-change-refugees-2100.
4. Diane Udel, "Seventy Percent of Florida's Coral Reefs Are Eroding, New Study Finds," *News and Events*, Rosenstiel School of Marine and Atmospheric Science (December 5, 2022) https://news.miami.edu/rosenstiel/stories/2022/12/seventy-percent-of-floridas-coral-reefs-are-eroding-new-study-finds.html.
5. Erica Towle et al., *Coral Reef Condition: A Status Report for Florida's Coral Reef* (NOAA Coral Reef Conservation Program, 2020), https://www.coris.noaa.gov/monitoring/status_report/docs/FL_508_compliant.pdf.
6. Regenesis Group, *Regenerative Development and Design: A Framework for Evolving Sustainability* (Wiley, 2016).

Chapter 1

1. Ethan Kent, "Reconnecting People through Places: Placemaking with Local, and Global, Networks," PlacemakingX (July 2024) https://www.placemakingx.org/article/reconnecting-people-through-places.
2. M. Kat Anderson, *Tending the Wild: Native American Knowledge and the Management of California's Natural Resources* (University of California Press, 2005).
3. Suzanne Dallman and Thomas Piechota, *Stormwater: Asset Not Liability* (Los Angeles and San Gabriel Rivers Watershed Council, 2010).

4. René Dubos, *A God Within* (Scribner, 1972).
5. Regenesis Group, *Regenerative Development and Design: A Framework for Evolving Sustainability* (Wiley, 2016).
6. John K. Grande, "Newton & Helen Mayer Harrison: How Big Is Here?" Espace Sculpture 101 (2012), 26–27.
7. Kevin Kelly, *The Inevitable: Understanding the 12 Technological Forces That Will Shape Our Future* (Viking, 2016).
8. United Nations Department of Economic and Social Affairs, "68% of the World Population Projected to Live in Urban Areas by 2050, Says UN," *News*, United Nations (May 16, 2018) https://www.un.org/development/desa/en/news/population/2018-revision-of-world-urbanization-prospects.html.

Chapter 2

1. US Climate Resilience Toolkit, "Seven Generations: Community-Based Environmental Planning," US Climate Resilience Toolkit, 2024, https://toolkit.climate.gov/tool/seven-generations%E2%80%94community-based-environmental-planning.
2. Helena Norberg-Hodge, *Ancient Futures: Learning from Ladakh* (Sierra Club, 1991).
3. Don Miguel Ruiz, *The Four Agreements: A Practical Guide to Personal Freedom* (Amber-Allen, 1997).
4. Florida Memory, "Photo Exhibits: Great Jacksonville Fire of 1901," State Archives of Florida (2024) https://www.floridamemory.com/learn/exhibits/photo_exhibits/jacksonvillefire.
5. The Regenesis Group, "The Regenerative Practitioner Series," The Regenesis Group (2020) https://regenesisgroup.com/the-regenerative-practitioner-working-program-overview.
6. Dominique Hes, "Changing the Way We Think about Ecocities," Pursuit (July 14, 2017), https://pursuit.unimelb.edu.au/articles/changing-the-way-we-think-about-ecocities.
7. Alexandra J. Tohme, (n.d.). "Rapid Growth Is Happening," Future of Cities, https://focities.com/author/ajtohme/.

Chapter 3

1. Joseph P. Lash, *Helen and Teacher: The Story of Helen Keller and Anne Sullivan Macy* (Delacorte Press, 1980).
2. Lee Pfeiffer, "Pinocchio," *Britannica* (January 16, 2025), https://www.britannica.com/topic/Pinocchio-film-1940.

3. Meg MacIver, "The Social Life of Our Urban Spaces," *Gapers Block* (November 17, 2008) http://gapersblock.com/detour/the_social_life_of_our_urban_spaces.
4. Dan Buettner, *The Blue Zones: Lessons for Living Longer from the People Who've Lived the Longest* (National Geographic, 2009).
5. Tyson Yunkaporta, *Sand Talk: How Indigenous Thinking Can Save the World* (HarperOne, 2019).
6. Jaime Lerner, *Urban Acupuncture: Celebrating Pinpricks of Change that Enrich City Life* (Island Press, 2014).
7. Dunn History, "Lemon City," Dunn History: A History of Florida through Black Eyes (October 23, 2001) https://dunnhistory.com/black-miami/lemon-city.
8. Alexandra J. Tohme, (n.d.). "Rapid Growth Is Happening," Future of Cities, https://focities.com/author/ajtohme/.
9. Emily Moody, "Artists Bring an Outdoor Gallery to Life in 48 Hours at the PHX-JAX Arts + Innovation District," Future of Cities (October 16, 2023), https://focities.com/artists-bring-an-outdoor-gallery-to-life-in-48-hours-at-the-phx-jax-arts-innovation-district/.
10. Jalal ad-Din Muhammad ar-Rumi, "Yesterday I was clever, so I wanted to change the world. Today I am wise, so I am changing myself," Goodreads (December 8, 2012) https://www.goodreads.com/quotes/551027-yesterday-i-was-clever-so-i-wanted-to-change-the.

Chapter 4

1. Gary Snyder, *The Practice of the Wild* (Counterpoint, 1990).
2. Alan W. Watts, *Cloud-Hidden, Whereabouts Unknown* (Knopf Doubleday, 1973).
3. Pamela Mang and Ben Haggard, *Regenerative Development and Design: A Framework for Evolving Sustainability* (Wiley, 2016).
4. Carol Sanford, *The Regenerative Business: Redesign Work, Cultivate Human Potential, Achieve Extraordinary Outcomes* (Elevate, 2017).
5. R. E. A. Almond, M. Grooten, and T. Petersen, eds., Living Planet Report 2020: Bending the Curve of Biodiversity Loss (WWF 2020), https://www.wwf.org.uk/sites/default/files/2020-09/LPR20_Full_report.pdf.
6. Ibid.
7. Michelle Marie Esposito, Sara Turku, Leora Lehrfield, and Ayat Shoman, "The Impact of Human Activities on Zoonotic Infection Transmissions," *Animals* 15 (2023): 1646, https://www.ncbi.nlm.nih.gov/pmc/articles/PMC10215220/#:~:text=Zoonotic%20diseases%20that%20likely%20originated,73%2C74%2C75%5D.
8. https://www.chozenretreat.com.

9. World Economic Forum, "Wetlands, the Forgotten Carbon Sink that Can Help Mitigate Impact of Climate Change," World Economic Forum (December 21, 2023), https://www.weforum.org/agenda/2023/12/wetlands-carbon-sink-climate-change-mitigation.

10. NASA Earth Observatory, "Climate Q&A: If Earth Has Warmed and Cooled throughout History, What Makes Scientists Think that Humans Are Causing Global Warming Now?," Earth Observatory (May 8, 2010), https://earthobservatory.nasa.gov/blogs/climateqa/if-earth-has-warmed-and-cooled-throughout-history-what-makes-scientists-think-that-humans-are-causing-global-warming-now.

11. Intergovernmental Panel on Climate Change (IPCC), *Climate Change 2021: The Physical Science Basis* (Cambridge University Press, 2021), https://www.ipcc.ch/report/ar6/wg1.

12. Roger S. Ulrich, Robert F. Simons, Barbara D. Losito, Evelyn Fiorito, Mark A. Miles, and Michael Zelson, "Stress Recovery During Exposure to Natural and

13. Urban Environments," *Journal of Environmental Psychology* 11 (1991): 201–230, https://www.sciencedirect.com/science/article/abs/pii/S0272494405801847.

14. Gregory N. Bratman, J. Paul Hamilton, and Gretchen C. Daily, "The Impacts of Nature Experience on Human Cognitive Function and Mental Health," *Annals of the New York Academy of Sciences* 1249 (2015): 118–136, https://nyaspubs.onlinelibrary.wiley.com/doi/10.1111/j.1749-6632.2011.06400.x.

15. Margaret M. Hansen, Reo Jones, and Kirsten Tocchini, "Shinrin-Yoku (Forest Bathing) and Nature Therapy: A State-of-the-Art Review," *International Journal of Environmental Research and Public Health* 14 (2017): 851.

16. James F. Sallis et al., "Physical Activity in Relation to Urban Environments in 14 Cities Worldwide: A Cross-Sectional Study," *Lancet* 387(2016): 2207–2217, https://doi.org/10.1016/S0140-6736(15)01284-2.

17. Ibid.

18. Aleksandar Radujkovic, Theresa Hippchen, Shilpa Tiwari-Heckler, Saida Dreher, Monica Boxberger, and Uta Merle, "Vitamin D Deficiency and Outcome of COVID-19 Patients," *Nutrients* 12 (2020): 2757, https://doi.org/10.3390/nu12092757.

19. Thomas Astell-Burt and Xiaqi Feng, "Association of Urban Green Space with Mental Health and General Health Among Adults in Australia," *Environmental Research* 180 (2019): 108–116, https://doi.org/10.1016/j.envres.2019.108-116.

20. Sathya Swarup Aithal, Ishaan Sachdeva, Om P. Kurmi, "Air Quality and Respiratory Health in Children," *Breathe* 19 (2023): 230040, https://pmc.ncbi.nlm.nih.gov/articles/PMC10292770/?.

21. Jon Ungoed-Thomas and Maximilian Jenz, "Fifty-Seven Swimmers Fall Sick and Get Diarrhoea at World Triathlon Championship in Sunderland," *Guardian* (August 5,

2024), https://www.theguardian.com/environment/2023/aug/05/investigation-after-57-world-triathlon-championship-swimmers-fall-sick-and-get-diarrhoea-in-sunderland-race.

22. Q Li et al., "A Forest Bathing Trip Increases Human Natural Killer Activity and Expression of Anti-cancer Proteins in Female (and Male) Subjects," *Journal of Biological Regulators and Homeostatic Agents* 22 (2008): 45–55, https://pubmed.ncbi.nlm.nih.gov/18394317.
23. Pamela Mang and Ben Haggard, *Regenerative Development and Design: A Framework for Evolving Sustainability* (Wiley, 2016).
24. Elisabet Sahtouris, *EarthDance: Living Systems in Evolution* (iUniverse, 1999).
25. Wilhelm Barthlott and Christoph Neinhuis, "Purity of the Sacred Lotus, or Escape from Contamination in Biological Surfaces," *Planta* 202 (1997): 1–8, https://link.springer.com/article/10.1007/s004250050096.
26. BBC News, "How a Kingfisher Helped Reshape Japan's Bullet Train," YouTube (March 30, 2019), https://www.youtube.com/watch?v=YVU6YBPaaB8&t=2s.
27. Never Enough Architecture, "The Eastgate Center," Never Enough Architecture (November 30, 2023), https://neverenougharchitecture.com/project/the-eastgate-centre.
28. Janine M. Benyus, *Biomimicry: Innovation Inspired by Nature* (HarperCollins, 1997).
29. Friends of the High Line, "The High Line," Friends of the High Line (2024), https://www.thehighline.org.
30. Eric Baldwin, "CopenHill: The Story of BIG's Iconic Waste-to-Energy Plant," ArchDaily (October 7, 2019), https://www.archdaily.com/925966/copenhill-the-story-of-bigs-iconic-waste-to-energy-plant.
31. Battersea Power Station Development Company, "About Battersea Power Station," Battersea Power Station (2024), https://batterseapowerstation.co.uk/about.
32. SUGi, "Greening Cities and Reimagining Urban Life," SUGi (2024), https://www.sugiproject.com.
33. SUGi, "The Miyawaki Method for Creating Forests," SUGi (July 15, 2024), https://www.sugiproject.com/blog/miyawaki-method-for-creating-forests.
34. SUGi, LinkedIn, https://www.linkedin.com/posts/sugiproject_sugi-pocketforest-biodiversity-activity-7187462928562872320-DBvU.
35. Dong Wang, Pei-Yuan Xu, Bo-Wen An, and Qiu-Ping Guo, "Urban Green Infrastructure: Bridging Biodiversity Conservation and Sustainable Urban Development through Adaptive Management Approach," *Frontiers in Ecology and Evolution* 12 (2024): 1440477, https://www.frontiersin.org/journals/ecology-and-evolution/articles/10.3389/fevo.2024.1440477/full.
36. Peter Yeung, "How a Colombian City Cooled Dramatically in Just Three Years," Reasons to Be Cheerful (March 4, 2024), https://reasonstobecheerful.world/green-corridors-medellin-colombia-urban-heat.

37. Big Green, "Project Overview: Million Gardens," Big Green (2024), https://biggreen.org/milliongardens.
38. ChoZen Retreat Center, "A Camp for the Humanity of the Future," ChoZen (March 31, 2025), https://www.chozenretreat.com.
39. Beacon Food Forest (2024), https://www.beaconfoodforest.org.
40. Ryland Engelhart and Josh Tickell, directors, *Kiss the Ground* (Big Picture Ranch, 2020).
41. Groundwork Jacksonville, "Emerald Trail," Groundwork Jacksonville (2024), https://www.groundworkjacksonville.org/emerald-trail.
42. US Department of Energy, "People Powered: Chris Castro, Director of Sustainability (Direct Current - An Energy.gov Podcast)," YouTube (August 15, 2022), https://www.youtube.com/watch?v=InDmig9U5uY.
43. Corinne Le Quéré et al., "Temporary Reduction in Daily Global CO_2 Emissions during the COVID-19 Forced Confinement," *Nature Climate Change* 10 (2020): 647–653, https://doi.org/10.1038/s41558-020-0797-x.
44. Emily Healey, Jon Traunfeld, Rachel Rosenberg Goldstein, *The Impact of Gardening on Mental Health* (University of Maryland, 2025).
45. K. Kaufman, "A Small Town in Ohio Creates Industry Buzz with Solar Plus Storage," SeaPower.org (May 12, 2016), https://sepapower.org/knowledge/a-small-town-in-ohio-creates-industry-buzz-with-solar-plus-storage/?.

Chapter 5

1. Jawaharlal Nehru, *The Discovery of India* (Meridian Books, 1946).
2. Milton Friedman, *Capitalism and Freedom* (University of Chicago Press, 1962).
3. Virginia Gorlinski, "Fado," *Britannica* (May 29, 2025), https://www.britannica.com/art/fado.
4. Italian Language Centre, "The Fine Art of Italian Hand Gestures," Italian Language Centre (September 23, 2021), https://www.italianlanguagecentre.org/the-fine-art-of-italian-hand-gestures.
5. Whitney Smith, "Flag of Japan," *Britannica* (February 7, 2025), https://www.britannica.com/topic/flag-of-Japan.
6. Nancy Hargrove, *The Renaissance: A Short History* (Modern Library, 2004).
7. Judith Lorber, *Paradoxes of Gender* (Yale University Press, 1994).
8. Rachel Klein, *The Essential Jewish Cookbook: 100 Traditional Recipes for a Flavorful Year* (HarperCollins, 2021).
9. Lisa Lee, *Chinese Cooking for the American Kitchen* (Tuttle Publishing, 2006).
10. Andrew Gordon, *A Modern History of Japan: From Tokugawa Times to the Present* (Oxford University Press, 2009).

11. Mikael S. Andersen, *Environmental Policy in Scandinavia: From the 1960s to the Present* (Routledge, 2018).
12. Sejong, "The Importance of Family Values in South Korean Culture," Sejong.com.sg (April 18, 2023), https://www.sejong.com.sg/the-importance-of-family-values-in-south-korean-culture/?.
13. John W. Berry, "Immigration, Acculturation, and Adaptation," *Applied Psychology* 46 (1997): 5–34.
14. Ronald Cohen, *Global Diasporas: An Introduction* (University of Washington Press, 2008).
15. Federico Acevedo, "Florida State Parks Threatened by Development," Florida Wildlife Federation (October 18, 2024) https://floridawildlifefederation.org/florida-state-parks-threatened-by-development/.
16. Carl Lisciandrello, "Floridians Balk at a Plan to Add Golf, Pickleball and Hotels to State Parks," WUSF (August 23, 2024), https://www.wusf.org/environment/2024-08-23/floridians-balk-plan-add-golf-pickleball-hotels-florida-state-parks.
17. Amos Chapple and Arben Hoti, "Culture Wars: The Fate of Kosovo's Churches," RadioFreeEurope (August 16, 2024), https://www.rferl.org/a/kosovo-churches-serbs-destroyed/33079905.html.
18. Janine Ungvarsky, "Scandinavian Design," EBSCO.com (2023), https://www.ebsco.com/research-starters/visual-arts/scandinavian-design?.
19. Richard Higgins, "Preserving Tradition and Embracing Modernity: The Case of Kyoto's Machiya Houses," *Architectural Journal* 232 (2020): 54–65.
20. "About 5." Grassfed Culture Hospitality, n.d. https://www.grassfedculture.com/about-us.
21. Tigre Sounds (@tigresounds)."Discover ZeyZey Miami: Carving a New Miami Soundscape." Instagram, May 16, 2025. https://www.instagram.com/p/DJuruS_uApz/?img_index=1.
22. Botmas, https://botmas.nl/.
23. Louis Tay and Ed Diener, "Needs and Subjective Well-Being around the World," *Journal of Personality and Social Psychology* 101 (2011): 354–365.
24. Anny Shaw and Hannah McGivern, "Special Report: Funding Cuts and Weak Economy Send UK's Visual Arts into Crisis," *Art Newspaper* (July 11, 2023), https://www.theartnewspaper.com/2023/07/11/special-report-funding-cuts-and-weak-economy-send-uks-visual-arts-into-crisis.
25. Wallace Ludel, "The Arts Made Up More Than $1 Trillion of the U.S. Economy in 2021," Art Newspaper (March 16, 2023), https://www.theartnewspaper.com/2023/03/16/arts-1billion-us-economy-2021-nea-report.
26. Americans for the Arts, https://www.americansforthearts.org/by-program/reports-and-data/research-studies-publications/arts-economic-prosperity-5.

27. Paul Laster, "How the Wynwood Walls Have Shaped Miami's Art Scene," *Architectural Digest* (October 3, 2019), https://www.architecturaldigest.com/story/wynwood-walls-have-shaped-miamis-art-scene.
28. Katie Wells, "Embracing Hygge and European Traditions of Coziness," *Stars and Stripes Europe* (October 22, 2024), https://europe.stripes.com/lifestyle/embracing-hygge-and-european-traditions-of-coziness.html.
29. Wikipedia, "Muhallebi," *Wikipedia* (February 18, 2025), https://en.wikipedia.org/wiki/Muhallebi.
30. Parliament of Australia, *The Koala—Saving Our National Icon* (Commonwealth of Australia, 2011), https://www.aph.gov.au/Parliamentary_Business/Committees/Senate/Environment_and_Communications/Completed_inquiries/2010-13/koalas/report/index?.
31. US Fish and Wildlife Service (FWS), "Bald Eagle Factsheet," FWS (March 24, 2021), https://fws.gov/media/bald-eagle-fact-sheet.
32. Japan-Guide.com, "Cherry Blossoms," Japan-Guide.com (March 7, 2025), https://www.japan-guide.com/e/e2011.html.
33. Bill Reed, "Shifting from 'Sustainability' to Regeneration," *Building Research and Information* 35 (2007): 674–680, https://www.tandfonline.com/doi/full/10.1080/09613210701475753#d1e213.

Chapter 6

1. Helena Norberg-Hodge, *Local Is Our Future: Steps to an Economics of Happiness* (Local Futures, 2011).
2. Alexander Robinson and Myvonwynn Hopton, "Cheonggyecheon Stream Restoration Project," Landscape Performance Series (Landscape Architecture Foundation, 2011), https://doi.org/10.31353/cs0140.
3. Rivers by Design, *Cheonggyecheon Restoration Project* (European Centre for River Restoration, 2023), https://www.ecrr.org/Portals/27/Cheonggyecheon%20case%20study.pdf.
4. Ibid.
5. Alexander Robinson and Myvonwynn Hopton, "Cheonggyecheon Stream Restoration Project," Landscape Performance Series (Landscape Architecture Foundation, 2011), https://doi.org/10.31353/cs0140.
6. Rivers by Design, *Cheonggyecheon Restoration Project* (European Centre for River Restoration, 2023), https://www.ecrr.org/Portals/27/Cheonggyecheon%20case%20study.pdf.
7. American Independent Business Alliance, "The Local Multiplier Effect: How Independent Locally Owned Businesses Help Your Community Thrive," AIBA (February 16, 2025), https://amiba.net/local-multiplier.

8. Colin Marshall, "Story of Cities #50: The Reclaimed Stream Bringing Life to the Heart of Seoul," *Guardian* (May 25, 2016), www.theguardian.com/cities/2016/may/25/story-cities-reclaimed-stream-heart-seoul-cheonggyecheon.
9. Borgen Project, "Successful Startup Kubik Expands Affordable Housing in Ethiopia," Borgen Project (July 27, 2023), https://borgenproject.org/affordable-housing-in-ethiopia.
10. Tom Jackson, "Ethiopian Low-Carbon Building Startup Kubik Plans Pan-African Expansion Post-Funding," *Disrupt Africa* (May 10, 2024), https://disruptafrica.com/2024/05/10/ethiopian-low-carbon-building-startup-kubik-plans-pan-african-expansion-post-funding.
11. Borgen Project, "Successful Startup Kubik Expands Affordable Housing in Ethiopia," Borgen Project (July 27, 2023), https://borgenproject.org/affordable-housing-in-ethiopia.
12. Tom Jackson, "Ethiopian Low-Carbon Building Startup Kubik Plans Pan-African Expansion Post-Funding," Disrupt Africa (May 10, 2024), https://disruptafrica.com/2024/05/10/ethiopian-low-carbon-building-startup-kubik-plans-pan-african-expansion-post-funding.
13. Jaco Maritz, "Ethiopia: Converting Plastic Waste into Affordable Homes," *How We Made It in Africa* (July 17, 2023), https://www.howwemadeitinafrica.com/ethiopia-converting-plastic-waste-into-affordable-homes/155819.
14. Emma L. Teuten et al. "Transport and Release of Chemicals from Plastics to the Environment and to Wildlife," *Philosophical Transactions of the Royal Society B* 364 (2009): 2027–2045, https://doi.org/10.1098/rstb.2008.0284.
15. Esin Kasapoğlu, *Polymer-Based Building Materials: Effects of Quality on Durability*, Irbnet.de (June 27, 2022), https://www.irbnet.de/daten/iconda/CIB_DC25697.pdf.

Chapter 7

1. Smart Growth America, "Complete Streets," Smart Growth America (March 7, 2025), https://smartgrowthamerica.org/what-are-complete-streets.
2. Salome Gongadze and Anne Maassen, "Paris' Vision for a '15-Minute City' Sparks a Global Movement," World Resources Institute (January 25, 2023), https://www.wri.org/insights/paris-15-minute-city.

Conclusion

1. Dorothy Gale, *The Wizard of Oz*, directed by Victor Fleming (Metro-Goldwyn-Mayer, 1939).
2. Pamela Mang and Ben Haggard, *Regenerative Development and Design: A Framework for Evolving Sustainability* (Wiley, 2016).

3. The Global Goals, "The 17 Goals," GlobalGoals.org (March 7, 2025), https://www.globalgoals.org/goals.
4. Dominique Hes and Chrisna du Plessis, *Designing for Hope: Pathways to Regenerative Sustainability* (Routledge, 2015).

Toolkit 2

1. Douglas W. Tallamy, *Bringing Nature Home: How You Can Sustain Wildlife with Native Plants* (Timber Press, 2009).

Toolkit 3

1. Quote Investigator, "Culture eats strategy for breakfast," Quote Investigator (May 23, 2017), https://quoteinvestigator.com/2017/05/23/culture-eats.

Toolkit 4

1. Project for Public Spaces, "William H. Whyte," Project for Public Spaces (January 3, 2010), https://www.pps.org/article/wwhyte.

Acknowledgments

Tony & Ximena Cho in front of the "Heart of Okeanos" prototype for The ReefLine at Miami's Art Basel 2022 event hosted at the Climate & Innovation HUB powered by Future of Cities® in Little Haiti—Miami, Florida

First and foremost, my appreciation goes to my incredible late wife and soul mate, Ximena Cho, for her unwavering support of my passion, for always believing in me and always encouraging me to believe in myself. She passed on June 17, 2025, at home, surrounded by love, community, and sacred ritual, during the writing of this book after an eight-year heroic battle with breast cancer. Her soft power and kind and compassionate nature inspired so many in our global community. Her passing has had such a positive and profound impact on our entire community that she has inspired me to dive deeply and to embrace death and dying also as a regenerative process, which has the potential to heal so many by removing fear and taboo around the most important and sacred event of all of our lives, death. Maybe my next book will be on regenerative dying.

I also want to thank all those who have taught me so much along the way—namely, Bill Reed and Dr. Dominique Hes, for embracing this book and contributing to it and encouraging me to get it published. And a big thanks to all those at the Regenesis Institute, including Pamela Mang and Joel Glanzberg.

Big thanks to Ethan and Fred Kent from PlacemakingX for leading the charge for so many years and being such a strong ally in this work.

I want to also thank Jacksonville's mayor, Donna Deegan, for her leadership and aligned vision in wanting to make Jacksonville the best it can absolutely be.

Thanks to Brian Kurz, who helped launch the Future of Cities Platform during the pandemic while living with us at ChoZen.

Thanks to Roxi Shashadee and to Amanda Joy Ravenhill for being early advisers for Future of Cities and our partners in the regenaissance.

Thank you to the incredible team at Future of Cities®, Amy Kynoch, Alexandre Tohme, Bruno Vitale, Brendan McKeon, Emily Moody, Alyona and Marc Assante, Logan Parris, and Michael Weil, without whom none of these demonstration projects or bodies of research would exist.

Thank you to Vishen Lakhiani for pushing me to elevate my game, to Dan Buettner for showing me the meaning of true friendship and supporting me selflessly, and to Peter Diamandis for inspiring me to think big and live exponentially.

Thank you to the entire team at Greenleaf Book Group, as well as Amy White, my editor and book collaborator.

And finally to all the incredible systems thinkers, changemakers, and visionaries who came before us, Buckminster Fuller, Janine Benyus, Helena Norberg-Hodge, Bill Mollison, David Holmgren, and so many others. We stand on their shoulders and my hope is that this next generation is the one that truly embodies these practices. May we live in a world that serves 100% of life.

About the Author

TONY CHO is a purpose-driven entrepreneur, visionary community builder, and leading voice in regenerative placemaking. A catalyst of Miami's urban renaissance, he played a key role in transforming the Wynwood Arts District and founded the Magic City Innovation District, helping position Miami as a global hub for creativity and innovation.

As the Founder of Future of Cities®, Tony continues to advance community-centered developments worldwide that activate creativity, restore ecosystems, and shape future-ready cities. Through education, storytelling, and hands-on demonstration of regenerative principles, he brings "ecosystems thinking" to life at projects such as the Climate & Innovation HUB in Miami, ChoZen Center for Regenerative Living in Sebastian, Florida, and the Phoenix Arts & Innovation District in Jacksonville, Florida.

Featured on Gaia TV's *Road to Utopia* and in outlets including *The New York Times* and *The Wall Street Journal*, Tony's mission is clear: to positively impact the lives of a billion people through innovation in the built environment.